# Cat Body,
## Cat Mind

# Dr. Michael W. Fox

DOG BODY, DOG MIND (2007) The Lyons Press, Guilford, Connecticut

KILLER FOODS: WHEN SCIENTISTS MANIPULATE GENES, BETTER IS NOT ALWAYS BEST (2004)

The Lyons Press, Guilford, Connecticut

THE HEALING TOUCH FOR CATS (2004) Newmarket Press, New York

THE HEALING TOUCH FOR DOGS (2004) Newmarket Press, New York

BRINGING LIFE TO ETHICS: GLOBAL BIOETHICS FOR A HUMANE SOCIETY (2001) State University of New

York Press, Albany, New York

CONCEPTS IN ETHOLOGY, ANIMAL BEHAVIOR AND BIOETHICS (revised edition, 1997) R. E. Krieger Publishing

Co., Malabar, Florida

EATING WITH CONSCIENCE: THE BIOETHICS OF FOOD (1997) NewSage Press, Troutdale, Oregon

THE BOUNDLESS CIRCLE (1996) Quest Books, Wheaton, Illinois

THE NEW ANIMAL DOCTOR'S ANSWER BOOK (1989) Newmarket Press, New York

THE NEW EDEN (1989) Lotus Press, Santa Fe, New Mexico

ST. FRANCIS OF ASSISI, ANIMALS, AND NATURE (1989) Center for Respect of Life and Environment,

Washington, D.C.

LABORATORY ANIMAL HUSBANDRY (1986) State University of New York Press, Albany, New York

THE WHISTLING HUNTERS (1984) State University of New York Press, Albany, New York

LOVE IS A HAPPY CAT (1982) Newmarket Press, New York

INTEGRATIVE DEVELOPMENT OF BRAIN AND BEHAVIOR IN THE DOG (1971) University of Chicago Press

# Cat Body,
## Cat Mind

Exploring Your Cat's
Consciousness and
Total Well-Being

## Dr. Michael W. Fox

**The Lyons Press**
**Guilford, Connecticut**

An imprint of The Globe Pequot Press

To buy books in quantity for corporate use
or incentives, call **(800) 962–0973**
or e-mail **premiums@GlobePequot.com.**

The Lyons Press is an imprint of The Globe Pequot Press.

10  9  8  7  6  5  4  3  2  1

Printed in the United States of America

Designed by Sheryl P. Kober

ISBN 978-1-59921-062-9

Library of Congress Cataloging-in-Publication data is available on file.

For Deanna, and all we embrace

# | Acknowledgments |

I wish to acknowledge the many cats, dogs, and other animals who have enriched my life and taught me much. I am especially grateful for the sensitive and skilled editorial work of Lilly Golden. To my publisher, The Lyons Press, a word of thanks for helping me inform and inspire more understanding and compassion in our relationships with cats and all creatures great and small.

# | About the Author |

Michael W. Fox was born and educated in the UK, earning his veterinary degree at the Royal Veterinary College, London, from where he graduated in 1962. His subsequent research into animal behavior and development in the U.S. earned a Ph.D. in Medicine, from London University in 1967.

After receiving the 1972 Washington University Distinguished Faculty Award, St. Louis, Missouri, where he continued behavioral and developmental studies in canids and cats, for which he earned a D.Sc. in animal behavior/ethology from London University in 1976, he chose to focus his knowledge and concerns on advocating for animal protection, rights, and environmental conservation, and in continuing his avocation as a teacher and public speaker.

Between 1967 and 2003 he served in various positions with the Humane Society of the United States, including Scientific Director and Vice President for Bioethics and Sustainable Agriculture. During this time he was a regular guest on Johnny Carson's *Tonight Show*, published two best-selling books, *Understanding Your Dog* and *Understanding Your Cat*, did wildlife and conservation work in Africa and India, and went on to author and edit over forty books for adults and children. He writes the widely read, nationally syndicated newspaper column "Animal Doctor" with United Features Syndicate, New York. His regular monthly animal column in *McCall's* magazine was the longest running column on animals in a U.S. magazine.

He has chaired and served on National Academy of Science committees, and travels the world as a spokesperson for animal rights and environmental protection.

Dr. Fox is a widely recognized expert, consultant, and lecturer on animal awareness, emotions, rights, and well-being; on human-nonhuman bonds and rights philosophy; on bioethics, biotechnology, humane, sustainable agriculture; and holistic health. His long-held basic premise that human health and well-being are inseparable from animal health and welfare and environmental protection and conservation is now gaining international recognition as a bioethical imperative and prerequisite for a viable future.

He currently lives in Minneapolis, Minnesota, with three good dogs, one from Jamaica and two from India, and his wife, Deanna Krantz, when one or both of them are not traveling to India and other parts of the world to address those issues that threaten the welfare and beauty of the natural world and cause our fellow creatures great and small so much pain and suffering.

# | Contents |

# | Introduction |

The primary purpose of this book is to enhance your understanding of your feline companion(s) and how you communicate with them. This book will help you to become more fluent in "cat talk" and know what your cat is feeling, intending, and wanting. It will then be easier for your cat to communicate with you—especially when your cat wants to play, is distressed about something and wants your attention, or is possibly ill.

The human-animal bond is not only strengthened through improved understanding and communication. It is also reinforced and affirmed through better care, and this book offers you precisely what is needed in this regard: a holistic approach to companion animal care and preventive medicine that will help guarantee a healthy and happy animal, and a happy and healthy bond between the two of you.

Many people are convinced that their cats are psychic and can read their minds. Others contend that their companion animals have helped them heal and have even communicated from beyond the grave. Is all of this sheer fiction? Or are we dealing with a set of phenomena that our affection-ate and accepting cats are revealing to us—a situation that calls for a radical change in our understanding of these animal companions, and indeed all creatures great and small? This book will help you find answers to these and other questions.

For many people—including children and especially the lonely and the elderly—one of the deepest and most significant relationships they have enjoyed has been with their companion animals. This fact has touched me deeply as a result of my consultations with thousands of individuals who have sought my advice as a veterinarian and animal behavior therapist. In this book, I have put together the knowledge, insight, and concerns from this service work for the welfare of animals to provide some basic measures, some of which are long overdue, to improve the health and well-being of cats here in the United States and abroad. After all, cats are very significant others for millions of people and countless families around the world.

In times past, when our ancestors lived closer to nature and to wild and domesticated animals, there was no doubt that animals spoke to humans and were recognized as sharing an inner wisdom and connection with the sacred or metaphysical realm. They became our guardians, teachers, and healers—provided we had an open heart and mind.

Today our hearts and minds are being opened by what amounts to a renaissance (literally, a rebirth) in our rediscovery of these animal powers and a recovery of the ancient wisdom, thanks to the millions of companion animals that share and enrich our lives as family members, playmates, and soul mates.

The only, if not the last, connection with all that is natural and not man-made for so many of us is our connection with domesticated animals. Their presence and spirit make some of us feel more whole and secure. For many, companion animals give unconditional relief from loneliness, alienation, depression, and despair. And for others, the animals in their lives are their only affirmation that there is something miraculous and wonderful, even spiritual, in the close relationship with another living being—and this can be a source of joy and deep satisfaction from childhood on.

Since our animal companions have been at the sides of mankind from generation to generation, they mirror in their health and well-being the sensitivity and compassion of the times. The ideal bond between us and them is one that is mutually enhancing and based as much upon love and respect as upon understanding. The better we understand the cat's mind and have a basic comprehension of how our cats communicate their needs, intentions, and feelings, the more mindful we will be and the more satisfying and meaningful the human-animal bond will be.

There are people who look down on other people's close bond with their feline companions, seeing the relationship as misguided sentimentalism or as some displaced parental urge. They are incredulous over how deeply such people mourn the loss of a beloved old cat, and they firmly believe that animals don't have emotions, self-consciousness, and souls. Companion animals are seen by such skeptics as being emotional parasites and, when compared to their wild counterparts, degenerate and inferior.

To regard companion cats as inferior to their wild lynx and bobcat cousins because they have been domesticated and made dependent on us is to reveal a prejudice based on ignorance rather than sound science. In fact, scientific studies have shown that while the brains of domesticated animals are slightly smaller than their wild counterparts (the same is apparently true for modern human brains compared to the skull/brain-case volume of Cro-Magnon ancestors!), their physiology and psychology have been profoundly influenced in a more positive way. They are less fearful, more trusting, more responsive, and more adaptable and trainable than their wild ancestors. But they have been made more vulnerable to us in the process.

This does not mean that wild animals, especially those born in captivity and raised by caring people, are incapable of establishing a close emotional bond with us. But rather, domestication has made the deep heart of companion animals more accessible to us. With less instinctual fear and more trust, cats can bare their souls to us and reveal, often to our astonishment and delight, some of their higher powers such as insight, reasoning, prescience, empathy, devotion, and emotional intelligence—traits and talents we thought were exclusive to only the best of our own kind.

Skeptics have argued that such revelations are all a projection of our own imaginations and of our need to feel loved and understood—and by so doing, we are guilty of anthropomorphizing animals, which means endowing them with human qualities that only we actually possess.

This book demolishes such skepticism, which leads ultimately to *speciesism*—which is regarding animals as an inferior race of beings—and it provides a wealth of evidence to the contrary. This book is a testimony to the higher powers of animals other than us, and it affirms that the cats that share and enrich our lives in so many ways are indeed self-conscious, sensitive souls who live more in the present than most of us. They have much to teach us about the nature of love and the love of nature—all that is natural and spontaneous—through our deep heart connection with them.

My friend and mentor the late Professor Konrad Lorenz, who received the Nobel Prize for his pioneering work in ethology, the science of animal behavior, once told an international gathering of scientists that, "Before you can really study and understand an animal you must first love it." Konrad would agree with me that he is half right, because before you can really love another human or nonhuman being, you must first have some understanding—and the greater the understanding, the deeper the love. Hence the first part of this book deals with what is so often lacking in the human-animal bond: not love so much as understanding animal behavior and communication.

As an ethologist and veterinarian who has shared his life with many animals—from dogs and cats to wolves, foxes, and coyotes (who do not make "pets")—I can attest to the fact that in the research setting of scientific study and in the veterinary hospital, animals are less likely and able to be themselves and to show their true colors, compared to when they are in their familiar and secure home environment with their loved ones. It is here, in their familiar surroundings, that a treasure trove of insight and understanding of their natures—their souls and their higher powers—can be found. Thanks to the consultations, lectures, and communications with people about their animal companions, I have accumulated a treasure trove from the deep heart of our feline and other animal companions that will now be opened and shared with all. I feel privileged to be able to contribute to the advancement of our understanding, respect, and appreciation of fellow creatures great and small and to join others in affirming and celebrating this wonderful bond.

In the back half of this book, I have included chapters that address the major companion animal health and welfare concerns. This book also provides essential information on holistic health and preventive medicine, nutrition, vaccination protocols, basic training, and dealing with some of the most common behavioral problems.

# Cat Body,
## Cat Mind

# Part One
## Cat Mind

# 1
## | The Case for Cat Consciousness |

### Cats as Teachers

Cats can be our teachers, allies, guides, and healers—when we are able to be open to them and observe their behavior and learn how to communicate with them. When we accept and embrace them as our teachers of animal consciousness and animal behavior, we come to understand and respect their many powers of perception and feeling, instinct, and intuition. As philosopher Arthur Schopenhauer wrote, "There is only one untruthful creature on earth, and that is man. All others are honest and upright in that they openly declare their nature and do not simulate emotions they do not have."

In American Indian and other traditional aboriginal cultures, animals are as much a part of their culture's spirituality and mythology as they are a part of the ecological community and social economy. In modern society, these animal powers and gifts are less appreciated and less understood—but they are still there, even in our highly domesticated animal companions like the cat. We should not take cats for granted or regard them as lacking the powers their wild cousins possess. Many of them have the ability to survive and multiply as feral animals and live in the wild by their own wits, quite independent of us.

Millions of individuals throughout the "civilized" world have one thing in common with aboriginal people: daily contact with one or more animals, usually cats and dogs. Fortunately, only a few keep wild animals—such as wolves, wild cats, and hawks—in captivity. Unlike cats and dogs, which have been domesticated for thousands of years, wild animals are unable to adapt in body, mind, and spirit to the domestic environment; they inevitably suffer

3

in captivity, including many wild cat–house cat hybrids like "ocecats" and "Bengals"—crosses respectively between ocelots and Asian leopard cats and domesticated cats.

## Acknowledging a Cat's Awareness

When we do not understand animals and fully respect them in their own right, whatever feeling we may have for them is, at best, sentimental affection. This is because our ability to empathize with them—to put ourselves in their place and imagine what they are feeling, thinking, desiring, and going to do next—will be extremely limited if our understanding of their behavior and psychology is rudimentary. In other words, the better we understand them, the better we can empathize with them.

Empathy is the bridge of compassion and of loving kindness to which cats are extremely sensitive and responsive. As our understanding develops, however, the nature of that love and the quality of our relationships with animals changes. They become even more a part of our lives, to a degree of satisfaction rarely equaled with our own species.

Those people who refuse to accept that cats and other animals have feelings tend to put them down by claiming that everything animals do is instinctive and automatic, and they aren't consciously aware of what they do and feel. In their eyes, animals are unfeeling automatons—mere biological machines. It is perhaps no coincidence that people who feel this way about animals often exploit them as a livelihood.

Yet there are many converts. One world-renowned surgeon who transplanted chimpanzee hearts into human patients (or were they human guinea pigs?) stopped participating in the practice when the screams of a female chimp touched his soul after he had taken her mate out of her cage to be killed for his heart. An American hunter put his gun down forever after he killed a Canada goose and saw her mate fly down to help and defend her as she struggled on the ground in the final death throes. A scientist I knew who regularly experimented on dogs and other animals in the laboratory decided to quit doing research on animals after his daughter brought home a stray dog, who the family adopted, and whose love and devotion—especially to the man—made him realize how sensitive and intelligent the animals he experimented on were.

I have heard of men on whaling boats sickened by what they do as they watch a harpooned whale die and drown while being helped by its companions, who sometimes ram the boat. What a contrast to whale researchers who can approach these gentle giants and even touch them while the whales swim under their frail little rubber dinghies, careful never to capsize the researchers' watercraft.

Those who deny that cats and all animals have feelings may be denying their own feelings, too. Animals do have feelings—and to believe otherwise is to deny our kinship with them. By so doing, we cease to empathize with them. In the process of denying animals their feelings, we become unfeeling toward them—and to be unfeeling is to be inhuman.

It is ironic how some research psychologists raise infant monkeys in total isolation as a model of separation anxiety in human infants, and how others give dogs and other animals inescapable electrical shocks, and then claim such treatment provides a useful model for studying anxiety and depression in humans. They don't feel there is anything morally wrong with what they do; some even claim animals don't suffer as a human would if subjected to similar treatment. Yet in the same breath, they claim that such animal studies are relevant to understanding human emotional problems. If animals were not emotionally similar to us, then such research would not be relevant.

Cats not only have feelings, they also have emotional problems that are very similar to our own. *Separation anxiety* is the term given by animal behavior therapists to the most common emotional problem in cats, which develops when they are left alone all day in the house. The fact that experts now recognize animals have feelings and emotional problems is paving the way toward a greater respect for animals and for their rights.

It is well documented that many zoo animals become depressed when their cage-mates die or are separated from them. Likewise, cats show depression—disinterest in food and in life itself—when a beloved owner or animal companion dies (see chapter 6), or when they are boarded while the owners are away on vacation.

\* \* \*

What follows is an overview of some of the basic aspects of animal behavior to help stimulate a greater understanding and appreciation of animals, so that we will be better able to empathize with them and be the recipients of their trust, affection, and myriad gifts and powers.

# Signs of Distress

Gross changes in behavior—such as flight, screaming, struggling, and defensive self-protective "fear-biting" rather than offensive aggression—won't always occur when a cat is distressed and suffering. One must be alert for more subtle reactions that may, in certain combinations and contexts, be clear indicators of distress. Here are some examples:

1. Changes in the autonomic nervous system, the unconscious homeo-static or balancing control system of bodily functions, are associated with physical and/or emotional distress. These include: salivation, pupil dilation, increased rate of heart beats (palpitations), respiration (panting), increased body temperature, muscular tremors (shivering) or muscular tension, piloerection (fluffed fur), urination, defecation, and release of anal gland secretions.

2. Disruption of normal cyclic and maintenance behaviors, such as not eating, drinking, sleeping, or grooming.

3. Disturbance in social and other behaviors is exhibited when the animal will not explore, play, interact socially—with its own kind or with humans (the animal may be passive, actively avoid, or show defensive aggression).

4. Abnormal behaviors may develop, such as prolonged refusal to eat leading to anorexia (wasting), polydipsia (excessive drinking), hyper-aggressive or flight reactions, excessive grooming, redirected (self-directed) aggression and self-mutilation, pacing and circling stereotyped movements, as well as increased susceptibility to disease.

*Displacement Behavior and Redirected Actions*
Many cats and other animals (including humans), when in a state of conflict or anxiety, may suddenly display a particular behavior that, to the attuned observer, gives a clear indication of the animal's emotional state. Self-directed scratching, licking, and grooming are especially common in a variety of mammals, from rats and cats to apes and people. Cats traumatized by being weaned too early will often suck on their own tails and paws, and on their human companions' fingers, arms, and ear lobes. Such actions are like a child's thumb-sucking and are thought to be self-comforting and there-

A depressed cat becomes withdrawn and seeks solitude.
Photo Robin Scott

A happy cat lies on her back, trusting and vulnerable.
Photo Aya Kinoshita

fore to reduce anxiety. But they can be symptomatic of obsessive-compulsive disorder. Displacement eating and drinking have been observed in cats and many other animals and are two amusing aspects of human behavior at a formal cocktail party or reception!

Some animals engage in displacement or *sham-sleep* when they are scared (like playing possum). Cats in shelters will sham-sleep as a way of coping with stress; to the naïve observer, they may simply seem relaxed.

Sometimes the behavior is not displacement but a redirection. Instead of attacking a rival, the cat may redirect the attack on a companion, against some inanimate object (such as a food bowl), or against itself (another cause of self-mutilation). Such redirected actions are common in many animals, including human beings. One such example is when a man or woman comes home after a bad day at work and turns that experience into anger toward the children or spouse.

*Stereotyped Behaviors*

You may have seen zoo animals (especially big cats confined in small cages) and some house cats in shelters and catteries pacing and circling their cages. These are called *stereotyped behaviors* and they will develop when an animal (or human being) is frustrated, anxious, hyper-aroused, or wholly understimu-lated in a sterile cage or prison cell. The movements are self-stimulating and may afford a kind of sensory escape from confinement. Sometimes they are self-comforting, just like an anxious or over-aroused autistic child or adult schizophrenic who rocks back and forth and either sucks a thumb or "self-clings" with both arms wrapped around the body.

Stereotyped behavior indicates that something is wrong. Recent studies have shown that this type of behavior results in increased production of natural opiates in the animal's body; this may help the animal cope with stress and distress, but it can make the behaviors addictively obsessive-compulsive.

## What Makes My Pet Happy?

This is the title of a brochure published in 2006 by the British Veterinary Association's Animal Welfare Foundation (www.bva-awf.org.uk). This land-mark publication acknowledges that happiness, welfare, and quality of life have to do with how animals feel. And that the first thing to do to determine

A stressed, fearful cat crouches, twitches her tail, has enlarged pupils, then defensively arches her back, makes a mock threat-attack, then flattens her ears and crouches.
Photos M.W.Fox

if your animal is happy is to check how you are treating your animal companion against the so-called Five Freedoms, which include both physical and mental health considerations.

These Five Freedoms are:

1. Freedom from hunger and thirst

2. Freedom from pain, injury, and disease

3. Freedom from discomfort (e.g., temperature extremes, uncomfortable floor surfaces)

4. Freedom to express normal behaviors

5. Freedom from fear and distress

Now most people are not remiss in meeting the happiness criteria of numbers one through three, by providing their animal companions with proper physical care and veterinary attention as needed. But numbers four and five may be more difficult to provide without some understanding of animal behavior.

There is much in this book that will help guide people who want their animal companions to be happy and healthy. But when in doubt about numbers four and five, when it comes to you having any misgivings about your animal's quality of life under your care that is a function of freedom from fear and distress and freedom to express normal behaviors (and I would add freedom from abnormal behaviors that can develop with improper rearing, handling, and training—or lack thereof), then always seek professional advice.

An animal's happiness and quality of life are profoundly influenced by the quality time spent with that animal every day—engaging in activities the animal enjoys, especially various games, being groomed, being massaged, and having the opportunity to live with or meet other members of the same species for social interaction.

# Letting You Know

Animals let us know when they are suffering and in pain in ways that are some-times obvious, and in ways that can be ambiguous. Some cats in pain meow loudly to solicit attention. Others become silent and withdrawn, masking their symptoms, or they seem to undergo a personality change by becoming more irritable, hissing, or even biting and scratching when touched.

Some cats simply go off to be alone when they are ill, which may be their way of healing by finding a quiet and secure place to let natural healing processes work. This behavior probably underlies the myth that cats have nine lives. A cat seemingly at death's door disappears for a while then comes home healed. But this may not always be the best remedy, and any animal who behaves in such a way should be taken for a veterinary checkup. Don't allow him outdoors until you've taken him to the vet because once outside, he may then be hard to find. This reclusive, self-cloistering behavior probably underlies the widespread observation that many cats who are ill and "know their time has come" go off alone to die.

Cats can be good communicators when they are sick and in need of veterinary attention. Elizabeth Grace of Old Chatham, New York, wrote to me about a stray cat who appeared at the door of her two-cat household, limping and holding up an abscessed front paw. "Although previously not handled by me, he allowed me to pick him up and take him to the vet for treatment. Needless to say he became our third cat and is still alive today."

Ellen Bowie of Washington, D.C., wrote: "It was late at night and I was asleep. Munchkin [her cat] jumped up on the bed, grabbed my arm with his paw and tapped on me until I was awake. Then he got so close to my face as he could and yowled. Of course, I knew something was wrong and got up . . . When he knew he had my attention he started walking all over the bed and squatting as if to demonstrate he couldn't urinate. As I got dressed to take him to the emergency clinic he kept pacing up and down the hall and running in to see if I was ready to go. It was the *only* time in his life he willingly jumped into his carrier. We had a happy ending. The veterinarian at the emergency clinic was able to flush out his urethra and unblock him. Now that I am much better educated about natural food, water, and supplements, he is just fine."

Wild animals will sometimes seek out human help when they are suffering. One remarkable example was a bobcat that was severely infected and dying from porcupine quills in its face, mouth, and chest; the bobcat crawled over to two cross-country skiers and lay down in front of them. He was taken to a veterinarian, treated, and later released back into the wild.

Another instance was recounted to me by a woman who cared for a sick and starving bobcat who came to her country-house porch one cold winter day. The wild cat was released after it had regained its strength and condition; some months later, the cat returned, just once, with three bobcat kittens following behind, "as though to express her gratitude to me," said the woman.

At the Animal Refuge in South India, which my wife Deanna Krantz and I operated for many years, we had several remarkable animal experiences. In two instances, animals in need of veterinary care brought themselves to us for treatment. One was a water buffalo with a severe screw-worm infestation that was eating her alive; the other was a dog who had a broken back and pelvis and dragged himself from where he had been hit by a vehicle on the road over a mile away. Neither of these animals had ever been near our refuge before, yet somehow they knew it was a place where animals were cared for and healed.

## Companionship for Animals

At our Animal Refuge, with over three hundred animals of all kinds, we found that some were very attentive to sick and injured animals brought in for treatment. It was touching to see a cow care for an orphaned foal with nuzzles and licks, to see a dog care for an injured fawn by providing body warmth, licks, nibbles, and a protective growl when any other resident animal came near, and to see an adult male monkey care for an abandoned baby monkey by holding her in his arms and cooing. These observations confirmed my long-standing belief that companion cats are happier and healthier when they live with at least one member of their own species in the home, or with another animal species that is compatible. The animals provide each other with companionship and tender loving care, especially when left alone in the home for extended periods during the work week.

# 2
## | Animal Awareness and Communication |

The fascinating work of behavioral scientists has recently hit the popular press. Their attempts to establish two-way communication with animals such as chimpanzees and dolphins have established conclusively that humans are not the only intelligent beings on earth. The old assumption is that nonhuman animals lack any conscious intent (or awareness) to communicate, while humans are the only species who know what they are doing when they communicate.

Some philosophers have argued that a creature cannot think if it has no verbal language and thus, since only humans have a verbal means of communicating feelings and intentions, only humans can think. This is a ludicrously anthropocentric view indeed—and one any cat owner would certainly object to. But this view has prevailed in science for many decades and is perhaps the kind of attitude that underlies so much of the animal abuse we see in today's society.

One of the reasons why the inner mental realm of cats and other animals hasn't been well researched is probably because few experimenters develop a close-enough rapport with their animals to be accepted by them and to become "at one" with them socially and emotionally. Cat owners who basically have the kind of relationship with their animals whereby the species barrier is broken should perhaps be emulated more by scientists of animal behavior.

Take, for instance, one of Dr. John Lilly's research dolphins: knowing which one of several colored balls to retrieve when given a verbal command by her trainer, the dolphin deliberately retrieved the wrong ball. Her retrievals were in reverese order, in a pattern that revealed she was communicating the strategy of the game or test. This is not unlike the cat who drops a toy at your feet for you to throw for her to retrieve, which she does; but the cat breaks the pattern by dropping it farther away from you, so you must now retrieve it for the cat. The cat knows the game, and is actually reversing roles making the person be the retriever.

## Body Language, Scents, and Sounds

Few people have tried or are inclined to mimic some of the behavior patterns of their feline companions. This is not quite as difficult as it may sound, once you become familiar with some of the animals' communication signals. I have described these at length in my book *Understanding Your Cat.*[1] If you try to mimic these, what very often happens is something akin to a revelation for the human, especially for a child, and sometimes no less for the animal. Suddenly a human being is communicating with the cat in his own "language" and consciously attempting to break the species barrier. Judging by the responses of receptive animals, this is a highly rewarding experience.

Cats not only respond to our tone of voice, but also to our body language. Stretching, often with eyes half closed, and yawning are signals of relaxation that cats pick up from each other and from us. Lying down on the floor and rolling over onto one side mimics a cat's invitation to approach and be petted or played with. Getting down on all fours and making eye contact then weaving your head side to side then gently batting one hand toward the cat is an invitation to play. Extending a finger toward the cat's nose to make light contact mimics the cat's friendly nose-to-nose touch with another cat. Pretending to dodge one way and then the other, or playing peek-a-boo around a wall or the edge of the sofa, can lead to a game of tag or catch-me-if-you-can, which humans and many other animal species play (and they often use similar invitation or intention signals, too).

---

1. Michael W. Fox, *Understanding Your Cat* (New York: St. Martin's Press, 1992).

Free roaming cats in Greyfriar's Church graveyard in Edinburgh. Friendly cats (possibly littermates) greet with nose-touch and head-rubs. A black cat slinks by, avoiding eye contact, but is confronted, frozen by a direct stare and then attacked and driven away.
Photos M. W. Fox

Flopping down together is an invitation to paw-play.
Photo Aya Kinoshita

We should also remember that in other animals, as with people, the same communication signal can be given in different contexts, but the sender and receiver are aware of the social context they share and don't misread each other—at least not all of the time. For example, while a direct stare can be threatening, avoiding eye contact in humans can be a submissive signal in one context, a rude cut-off by a social superior in another, and a sexual coquettish come-on in yet a third context. Cats are very eye-contact sensitive. Simply turning and looking briefly in one direction then turning back may mean, "Look over there," or "Follow me," or "My food bowl is empty." When humans do not pick up on these subtle nonverbal communication signals and are not conversant with felinese, I am sure cats get quite frustrated and wonder how stupid humans can be.

Animals also use the *same* signal in various contexts, which has different meaning according to which context it is given. A good example is the growl-yowl-hiss of a cat, which in one context is a threat, a warning in another context, and a mere sham during playful scrapping. What this means is that an animal's awareness of itself and others is far more developed and sophisticated than many human skeptics might think.

A nose-to-nose touch greeting between two cats is simulated by a finger-to-nose touch greeting between a stranger and a cat.
Photos M. W. Fox

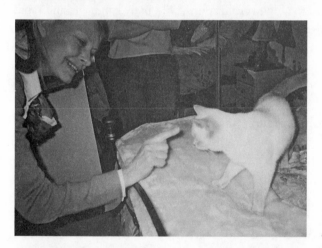

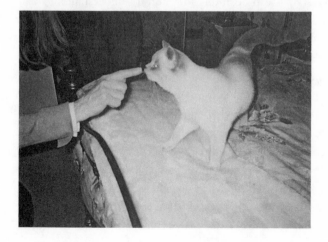

Sometimes, however, a person might misread you because he or she may not share the same context or set of expectations as you do. For example, a friendly laugh or smile may be mistaken for derisive laughter or a sneer; an admiring gaze might be mistaken as a rude stare or sexual invitation. This becomes a particular problem when the other person is paranoid or psychotic; the same applies to animals. They, too, having this context- and signal-related awareness, can get into misunderstandings. A playful, excited young cat with back arched and a fluffed-out tail may be misread by us or by an unfamiliar cat as readying to fight. A direct, curious but friendly stare from a person (especially one who is dark skinned and whose white eyes therefore create more of a contrast against a dark face) can be misread as a threat and intimidate a cat.

Cats not used to the loud vocalizations, wild body language, and rapid and unpredictable actions of children are often spooked and run off in terror—possibly interpreting such actions as threatening. I firmly believe this illustrates the importance of teaching children from an early age to be aware of animals and respect them by being quiet and moving slowly and carefully.

One particularly amusing "misreading" of signals was related to me by several cat owners whose felines reacted in bizarre ways to their sneezing. They told me their cats will always meow when someone sneezes, or yowl, make "chattering" sounds, and walk away, or puff up, wrinkle the nose, cackle, and roll over. One cat always yawns when his human companion yawns.

When animals or people do not share the same conceptual space in a given social context, signals can be ambiguous and misunderstandings will arise. This again indicates that animals are intelligent and aware, as does their ability to mirror or mimic some of our behaviors.

Perfume can be an ambiguous signal, especially for animals with a very sensitive sense of smell. Cats, for example, are sometimes clearly disturbed, sometimes even sexually aroused, by certain perfumes and the face and body creams that their owners put on. What signals might these owners really be giving to their cats one may only wonder. Cats have become almost psychotically afraid or aggressive when their owners put on certain expensive French perfumes that actually contain musk from civet cat anal glands!

Interestingly, cats have perfumed areas on their furry temples, cheeks, chins, lips, and along their tails, the precise function of which has yet to be determined. These body scent areas may play a role in social and territorial behavior, since cats rub these areas of their bodies on each other and against various objects in and around the home.

French scientists have isolated the pheromones from the skin scent glands of cats and from around the breast skin of nursing dogs and have found that dispensing these aromas to cats and dogs respectively can help reduce some behavioral problems, especially those associated with loneliness and separation anxiety. But better, in my mind, to get another animal rather than a comforting yet ghostly vapor!

By closely observing how our cats behave, we can quickly learn the rudiments of cat talk, or felinese. Stroking a cat is akin to one cat grooming another. Playfully (but gently) pawing her head or pretending to catch her tail (especially if she is playfully twitching it for you) are play actions one cat

Cats relish hide-and-seek and ambush games.
Photos Robin Scott

A typical facial expression of an alert, confident cat; closed eyes expression of a submissive, friendly cat.
Photos M. W. Fox

will make for another. We mimic many of these species-characteristic actions and signals with our cats, because they are part of our own repertoire and also because the cat "teaches" us (or shapes and reinforces our behavior) to use certain signals in preference to others by virtue of their appropriateness and effectiveness.

Even without further documentary evidence, it is obvious from what has been said that animals do have a very sophisticated sense of awareness and undoubtedly a much higher degree of consciousness than we might think. They know their own "language," elements of which are shared with other species, including us; they can learn, as in the case of a cat in a mixed-species family, to understand and sometimes mimic the communication of the other species. It is fascinating, for example, to observe a cat who has learned to relate to a dog, or to a pet rabbit.

A heightened sensitivity toward animals' signals will also be transferred to one's relationships with people, giving an added depth of awareness to the subtleties of human communication. Tension in the shoulders, nervous swallowing, fist clenching, self-clasping, and shifts in posture, eye contact, and gesture emphasis while speaking and being spoken to will be noted with greater clarity, affording more effective interpersonal relatedness through empathetic understanding.

The time is ripe for people to break through the species barrier and learn to relate more effectively with their animal companions by observing and mimicking their behavior. For too long perhaps, animals in our homes have sat around learning all our nuances, signals, and direct commands: let's equalize things and establish real two-way communication—and perhaps communion too!

## Understanding Cat Talk

It is important to recognize the signals that express a cat's emotions and intentions. For example, while a friendly dog will jump up on your lap and lick your face to express closeness, cats will simply sit quietly two or three feet away. Most dogs are at your feet when you speak to them, but cats will look at you from a distance, flick an ear or tail, and then close their eyes. Don't interpret this as meaning that the cat is lazy, aloof, or indifferent. It means just the opposite: it's the cat's way of stating, "I'm comfortable in your presence," and proving it by exaggeratedly relaxed gestures. The play and courtship signal of flopping over onto one side is a similar display of relaxation. Cats display submission not by rolling over onto one side like dogs, but by crouching, remaining still, and avoiding eye contact.

*Communicating Dominance and Submission*
Making and breaking eye contact is an intrinsic part of cat-to-cat interaction. A dominant cat will stare, and a subordinate will look away. Whenever you wish to discipline a cat verbally, make eye contact to assert your dominance. I call this the "face off" that is a pan-species communication signal exemplified by the dominant cat who looks through you and beyond. Any cat who comes into his or her field of vision almost invariably cowers and avoids direct eye contact.

Body postures displaying various emotional and motivational states: (a-b-c) increasingly offensive; (a-d-e) increasingly defensive/submissive; (a-f-g) increasingly intense defensive display.
Foxfiles

A cat that is afraid will have dilated pupils, while a dominant and aggressive one will have small pupils. A dominant cat generally acts very "cool" and does not arch its back or fluff itself into a Halloween cat display. It will walk slowly, keeping its eyes fixed on its rival, sometimes uttering a low growl and lashing its tail. A more fearful cat will arch its back and tail to give an illusion of greater size. It will hiss and, if very intimidated, it may scream.

Facial expressions in different emotional contexts. Note changes in ear position and pupil size: (a) alert; (b-c) increasing fear/apprehension; (d) ambivalent offensive-defensive expression, shifting to increasing defensive threat in e and f.
Foxfiles

*Showing Affection*

The friendly and affectionate signals that one cat gives to another are the same as those it gives to its owner. Approaching with the tail up vertically is an infantile signal, probably related to the kitten presenting its hind end to be cleaned by its mother. Raising the hind quarters while being petted may also be derived from this and should not be interpreted as sexual presentation.

A friendly cat will touch noses with another and rub its lips and head against its companion or nearby objects. These actions result in the companion (cat or person) being anointed with scent from special glands on the cat's lips, chin, and temples (the area in front of each ear on the head). The scent is not usually detectable by the human nose, but friendly cats mark each other in this way, the familiarity of shared odors being important in maintaining their social affiliation. A cat will rub its tail on furniture because there are scent glands on the tail, which, together with the lips and head glands, make the cat feel at home in a well-marked territory.

After briefly scent-rubbing a human or animal companion, a friendly cat will usually engage in social grooming. This is why cats so often lick their owners. Purring while grooming is a "stay together" signal. For that reason, we should respond to a cat's licks and purrs by grooming it: always stroking it along the lay of its hair, and talking in a soft voice.

The subtle feline greeting of vertical tail display followed by a head-rub, a nose-touch, and reaching up to paw.
Photos M. W. Fox

A kitten rolls over and stares, soliciting play; the adult flops over and they play-fight.
Photos Robin Scott

## Playfulness

Many cats become wilder at night—more active and playful and restless. This is a normal circadian activity cycle associated with nocturnal hunting and other nighttime activities. This is a good time for people to play with their cats.

*Playing with Your Cat*

Cats will often signal their readiness to play by approaching you and suddenly flopping onto one side. This is an invitation to you (or to another feline friend) to playfully come and tussle with the cat, or to hold a toy on a string above the cat to be caught, fought, and "killed." Cats will also arch their tails in an inverted U—a chase-and-catch-me-if-you-can display. Late evening is when they generally get these "cat crazies."

Cats of all ages enjoy toys and games. An adult cat "kills" a toy mouse.
Photo Aya Kinoshita

A kitten leaves its box "den" carrying its "prey."
Photos Robin Scott

All kittens, and some older cats too, will "hallucinate" during play, fluffing up and hissing, racing off as though something is chasing them, or leaping and batting at something between their paws while, in fact, nothing is there. Such bizarre behavior suggests that cats have vivid imaginations.

When cats hide behind furniture and then leap at your ankles, the cats are playing their own game of prey-catching; provided they don't scratch you, there's no need to discipline. A catnip-stuffed "mouse" or other suitable toy, like a sock tied to the end of a string attached to a short rod, is a good substitute for human ankles.

Mother cats will twitch the tips of their tails to provoke their kittens into playfully attacking their tails. We can mimic this by wriggling our fingers on the floor, on the place where we are sitting, under a bedsheet, or along the edge of the sofa.

Few cats will retrieve (Siamese cats being a significant exception), but many like to chase, catch, and "kill" long shoestrings or a catnip-filled sock that you pull to make it move like a mouse. They like to carry around various toys, such as socks and balls of crinkly paper. Be sure none of their toys could ever splinter if chewed or maybe swallowed.

Play is enjoyable and a physically and mentally beneficial activity that helps establish and maintain a strong social and emotional bond between participants. People who play with their cats will have happy and alert animal companions from their infancy into old age. Our animal companions also enjoy playing with each other, which is also fun to watch. Adopting two kittens from the same litter is a good choice since they are most likely to get on well and be good companions their entire lives.

*Tail Play*

The canine and feline species come naturally with a tail, a highly evolved, biologically adaptive organ of expression, communication, and intention. I once observed two Arctic foxes playing together. Their full bushy tails, which they would tuck their faces into to keep warm in the bitter winter tundra nights, played a major role in their play-fighting. Their "contact" sport was a ritualized dance: chasing, wrestling, leaping, and biting; gaping and gargle-screaming; panting, grinning, and bowing—with tails up vertically or arched over their backs.

One would rush at her partner and at the last moment twist and whisk her tail across his face. This was his cue to grab her tail—and only then would she take off and then literally have a tug-of-war wrestling match with her mate for her tail. They took turns, each using the other's tail as an object of play, a social tool-using behavior, where the tail was a catalyst for fox fun and games. A mother cat will tease her kittens, inciting them to attack the tip of her tail that she twitches to get them to respond—possibly to activate their natural hunting instinct and to learn self-control, because they will be disciplined if they bite too hard!

Some who first see cats at play think they are acting aggressively. One student who was house-sitting for me and caring for my two cats called me frantically on the phone asking what he should do because my cats fought each other every night. He described how they were behaving, and I assured him that this was their evening "crazies," during which time they would stalk each other, hide and ambush, race wildly over, under, and around furniture, and then engage in mock fighting by grabbing, biting, clawing, and growl-yowling at each other—but never did they ever harm each other.

## Intelligent Thinking and Insight

For those who believe animals lack the vestiges not only of human feeling but also of insight, reasoning, and foresight, the following examples of such capacities in cats and dogs should dispel such ignorance and prejudice.

- Insight. A cat sees her favorite toy behind the sofa but cannot reach it, then sees a string attached to the toy lying at the far end of the sofa. The cat seizes the string in her claws and quickly retrieves the toy.

- Reasoning. A cat observes its owner pressing a lever on an electric can opener. The cat starts to press the lever and operate the opener when she is hungry, signaling to her owner that it is mealtime.

- Foresight. A cat runs and hides when a cat-carrying cage is brought into the living room. The cat hasn't seen the cage for a year but correctly anticipates a trip to the veterinarian for his annual checkup.

A farmer told me about a tame coyote that he kept in a kennel on a long chain where it could do no damage. Over a period of time, a number of his chickens mysteriously disappeared, and he became suspicious. Hiding out near the barn, he saw the coyote take some of its own food and some corn cob (from a stored cache) and lay them out at the end of its chain. Then the coyote ran back into the kennel and waited in hiding for its prey. No human hunter using bait to lure could have shown more reasoned insight.

Author and naturalist Hope Ryden gave me a photo of a bobcat she saw fishing. The wild cat was crouched by the edge of a stream and was gently making the water ripple by batting the surface lightly with a paw, mimicking the actions of an insect in the water so as to lure fish within striking distance.

Do animals sit and wonder or worry about things like we do? That is difficult to know, since they can't speak; but it's quite clear when a cat is anxious or apprehensive. Cats will also have nightmares, crying and running in their sleep after a particularly traumatic day—such as one when the cat was attacked by another animal or had an upsetting trip to see the veterinarian.

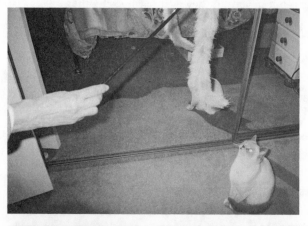

Cats enjoy games with humans, especially catching prey-like toys. Play has a bonding function.
Photos M. W. Fox

## Playful Insights

Child psychologists have learned much about the emotional state, imagination, and creativity of children from observing them at play with their dolls and other toys. Likewise, closely observing how one's cat plays with various toys can provide similar insights into the psyche. Here are some examples for you to interpret, followed by my own interpretations.

1. A cat carries a sock or another toy all over the house, meowing loudly, and then "hides" the object in a closet.

2. A cat adopts a fluffy slipper or toy a few weeks after she came out of heat, becomes very protective of it, and even sleeps with it.

3. A cat brings out all his toys when visitors come, chews, sucks, and kneads them vigorously but growls protectively when approached.

4. A cat brings a sock or another toy to you, places it at your feet or in your lap, and then walks away not soliciting you to throw it so it can be retrieved or "hunted and killed."

5. A cat brings a favorite toy to a houseguest, places it in the guest's lap or side, and looks with anticipation at the guest.

My interpretations follow.

1. The cat is "fantasizing" about carrying a kitten around.

2. The cat is having a false pregnancy and the slipper or toy is a kitten substitute.

3. The cat feels insecure and may be jealous of the attention being given to the visitors, or it may be afraid, needing the security of the toys.

4. The cat is giving you a gift—not unlike bringing food (prey) home for the kittens.

5. The cat is making friends, being a good ambassador and soliciting play and attention.

The more we observe our cats at play, and also play with them and give them suitable (and safe) toys to play with, the more they can communicate with us—and the more we can overcome the barrier of verbal language and learn about their psyches. The profound bonding function of play (animals who play together stay together) is critical for all who wish for a close relationship with their cats. Beginning at an early age, regular play sessions are as important as regular feeding, grooming, and exercise. Through play, animals learn to be gentle, develop self-control, and refine various actions associated in other contexts with fighting, prey-chasing, killing, and so forth. During social play and play with a toy, they also can show a sense of humor, creativity, and imagination, as well as joy in physical contact and various activities such as chasing and being chased, hiding and ambushing.

In his controversial book *The Question of Animal Awareness*, Professor Donald Griffin observes that, "Behavioral scientists have grown highly uncomfortable at the very thought of mental states or subjective qualities in animals. When they intrude on our scientific discourse, many of us feel sheepish, and when we find ourselves using such words as fear, pain, pleasure, or the like, we tend to shield our reductionist egos behind a respectability blanket of quotation marks."[2]

To doubt if an animal can experience pain, fear, anxiety, satisfaction, and pleasure is to doubt the very existence of our own consciousness. And to reject the possibility that our most recently evolved animal kin—the carnivores and primates—cannot or do not experience comparable states of joy, depression, guilt, remorse, and love is as illogical as denying that you or I have such experiences.

In the next chapter we will delve more deeply into cat behavior and communication, greatly improving our comprehension of felinese in the process, and look at behavioral problems and various psychological disorders that, with a little understanding, can be quickly remedied or prevented.

---

2. Donald Griffin, *The Question of Animal Awarenes* (New York: Rockefeller University Press, 1981).

# 3
# | Correcting Behavioral and Communication Problems |

In becoming integral members of the human family, cats learn to modify certain behaviors and acquire others to better communicate their basic social and emotional needs. In the process they "train" their human companions just as much—and sometimes more so—than they are trained by their owners. Many cats have trained their human companions when it is time to feed them, to wake up, and to play.

Through the interaction of genes (heredity) and early conditioning, cats develop different temperaments; and through the interaction of temperament with learning and social relationships, distinct personalities are formed.

Some people claim there are impossibly wild, aggressive, or fearful cats who can't be tamed—and that it is primarily human influences that determine how a cat's personality develops.

But genetic factors do influence temperament and certainly affect the quality of the bond we may enjoy with the animal—to the degree that no matter how much love and understanding is given during early, critical, and sensitive periods of development, the animal in question never really seems to fit in or enjoy life, just like some children.

## Natural and Bothersome, but Not Serious, Feline Behaviors

Some normal, instinctual behaviors of cats can cause owners concern. The rolling, rubbing, self-licking, and calling of an un-neutered female cat (especially a Siamese) are often misinterpreted as hysteria. They are, in fact, signs of sexual arousal. The cat is in heat and so pent-up that the courtship display of rolling and rubbing is spontaneous and may be directed toward humans in the home.

Urine is also used to mark territory, and males especially will spray certain objects to establish and proclaim their property rights. When this occurs indoors, the cat may be emotionally insecure (caused, for example, by the presence of visitors or the addition of a new human or animal member to the family) and thus in need of more attention. Or the cat may be frustrated, attention-seeking, disturbed by cats outdoors, or simply overly motivated. In the latter case, neutering will usually help reduce or eliminate the problem.

Cats need scratch posts. This one should be taller: cats like to go into a full stretch.
Photo M. W. Fox

When cats claw at carpets and furniture, they are not simply doing this to sharpen their claws. The action may be a scent-marking activity. It may also be associated with territorial marking; it may be a display of excitement, as when the cat's human companion comes home; it may be a signal that the cat is in the mood to play. Since this display generally occurs in specific places (near the door, for instance), setting up a scratching post or a log in that same location will protect your furniture.

"Love bites" can be disturbing to an owner—especially when cats holds on, with their eyes looking glazed or glassy as though they are in a trance. They are actually aroused sexually. A love bite may be followed by straddling an arm or leg, which the cat then "humps." This love bite is comparable to the scruff-bite of a male trying to hold its mate. If the bite isn't too hard, it should be accepted graciously and with humor, ditto the humping. A loud hiss and a tap on the nose should bring the cat back to reality and inhibit such behavior.

Cats are extremely sensitive to odors. A cat that comes home with a different smell (after a sojourn at the veterinary hospital or being out all night) may be attacked by a resident companion cat because it smells different. And, as noted, people have been attacked by their own cats when they wear a perfume containing musk that comes from the anal glands of wild civet cats.

Cats will frequently needle a person with their claws and make kneading movements with their paws while being petted. They will even drool a little at the mouth. This is quite natural social behavior. At such times, the cat is *regressing*—acting like a kitten and sham-nursing. Some cats like to nuzzle into a person's ear or armpit, which is an infantile social behavior that often occurs between cats. A mild tap on the nose will stop it. Kittens that have been weaned too early are more likely to develop the "vice" of nursing on blankets, chewing up woolens, and even sucking on their own tails as adults.

A cat that "loses face" after being disciplined will often very briefly lick its paw and then rub its face with that paw. This gesture indicates that the cat is embarrassed; it is not unlike a nervous person adjusting his hair or clothing. Excessive self-grooming can lead to self-mutilation.

## Scratching Posts

Every home should have a cat scratch post or board or a cat "gym" or "condo" with the scratching surface spiced with a little catnip. Some cats need to be shown how to use the scratch post by an owner making scratching movements with their fingernails or gently holding the cat up against the surface. Scratching posts should be secure, since a wobbly one will scare most cats. The post should be at least a few inches longer than the height of the cat when up on her hind legs and reaching as high as she can with her front paws. The surface should not be looped carpeting, which could snag the cat's claws.

Carpeted "cat condos" serve as good scratch posts.
Photos Robin Scott

## Litter Box Training

With a new kitten, carry her to the litter box and let her investigate. Take her back every time she has eaten and after she awakens from a nap. Place the litter tray in a quiet and easily accessible place. For kittens, the tray should be shallow. Remove feces and urine-soaked litter every day; clean the tray out and put in fresh litter as needed.

Cats naturally respond to the feel and texture of the kinds of litter box materials commercially available by sniffing, pawing, and then posturing to evacuate. Some cats will dig a small hole first, and most will cover their urine

The litter box should be in a quiet corner. Cats need peace and quiet when seeing to toilet functions.
Photos Robin Scott

or feces with litter after they have evacuated. A layer of baking soda on the bottom helps control odor.

It's best to stick to one good brand of litter. Switching brands often can make some cats refuse to use their litter boxes. Avoid buying litter that contains a lot of fine dust, which could cause problems for the cat. Some litter boxes have a removable cover. A covered litter box helps stop the cat from pawing the litter out of the tray. It is important to keep the litter box clean, because accumulated urine can make the interior of a covered litter box irritating to the cat.

## More Serious Cat Behavior Problems

*Lapses in Litter Box Training*

Breakdowns in toilet training are among the most common behavior problems in cats. If the litter tray is clean and there is no evidence of a clinical problem like the feline urologic syndrome (that includes cystitis, painful inflammation of the urinary bladder and urinary incontinence), constipation, or impacted anal glands and arthritis in old cats, especially affecting the spine (all of which can cause pain-associated aversion to the litter box), then an emotional reason should be looked for.

When cats feel insecure, they may stop their normal practice of burying their urine and feces. It is as though they must mark their home base even more with their own odors to feel safe. A new cat prowling outside, a new baby, or a visitor to the home can trigger this reaction. A change in relationships among cats living together can also be the cause, and then a new home will have to be found for the bully or the insecure house soiler.

Cats sometimes refuse to use the litter box because it is not kept sufficiently clean, especially when other cats share the same box. A change in the type of litter may also lead to litter box aversion. Different types of litter should be tried in three or four separate boxes. As mentioned before, boxes with covers that are not frequently cleaned can become filled with noxious fumes from the cat's urine, which is an extremely irritating and aversive situation for any cat. Dusty litter can cause respiratory irritation and litter box aversion too. Some cats like the box to be kept in a quiet, secure corner, or in a place that is not too dark, so the location of the litter box in the home must also be considered.

Where the cause of house soiling is anxiety related, your veterinarian might prescribe an anti-anxiety drug, like diazepam (up to 1 mg/day, orally), for three to four weeks. (Note: Do not clean soiled areas with ammonia, since this mimics cat urine and may encourage the cat to urinate on cleaned areas; a biological enzyme cleaner or dilute solution of white vinegar is preferable. Punishing the cat is likely to aggravate the problem.)

Once all possible causes have been checked and eliminated, the cat may have to be retrained to use his or her box. Keeping the cat in a small room or holding pen for a few days with a pillow, food, water, and its litter box close by will often help the cat to "reconnect." Time out of the confined area for play, petting, and brushing should be frequent, but these times should be strictly supervised so the cats aren't allowed to go near the places in the house where they have marked. Again, these spots should be shampooed or cleaned with white vinegar (diluted in equal parts water) or an enzyme cleaner and temporarily covered with a plastic sheet.

## Cats Who Spray Indiscriminately

A very common neighborhood complaint concerns owners' cats that are allowed to roam free outdoors and spray on neighbors' property, around their doors and window sills. Cats do this to mark their new territory and to mark that they've been there. Male cats tend to spray more than females, and tomcats spray more than neutered males.

## It's All in the Litter

Many people have contacted me about their concerns over the safety of various types of litter used in cat boxes. Dust, fine particulate matter in the litter that the cat may kick up while evacuating, causing lung irritation and respiratory problems, is a major issue, leading some cats to develop an aversion to using the litter box, and thus becoming house-soilers. Enclosed litter boxes, if not routinely cleaned, may become aversive to cats entering the enclosed space because of the build-up of ammonia.

There has been much speculation over the possible health risks to cats when so-called clumping litter, which aggregates into a ball where the cat urinates in the litter box, is used. And some cats are allergic to corn in their diets, so corn-husk cat litter may cause problems for cats who have developed a hypersensitivity to corn.

Ken Jones, DVM, of Santa Monica, California, shared with me his concerns and clinical observations on the risk to some cats of certain types of cat litter. He wrote:

> I have seen the occasional cat ingest cat litter when they are anemic, perhaps trying to compensate for an additional need for minerals in the diet. I have also seen some cats become impacted from litter when it gets stuck on their feet and they lick it off when cleaning themselves. I also have made an association a few times with certain types of litter contributing to bladder issues e.g. crystals and stones. The LUTD of crystals goes away when the litter is changed to newspaper, corn, wheat etc.
>
> Also, declawed cats may become agitated when using clay or pellets as this type irritates their feet due to the pads becoming small and hard.

Free roaming cats may use neighbors' flower beds and vegetable gardens as their toilets, which can be a health hazard because they may have harmful bacteria, roundworm eggs, and toxoplasma cysts in their feces that could infect people—children and pregnant women are especially at risk around cat-contaminated soil and sand boxes. When neighbor cats come around homes and spray everywhere, indoor cats smell the odor and may become upset and start fighting with each other and spraying in the house to mark and "defend" their territory. I once had a neutered Siamese cat who would back up to the window fan in the summer and spray into the fan to fill the room with his scent! Use either diluted white vinegar or an enzyme-based cleaner to deodorize sprayed areas.

It is best to start right by not allowing a kitten to get a taste for the outdoors for the many reasons detailed in Chapter 20. Cats adapt well to indoor living, especially if you have two cats to keep each other company. If you have a safe, dog-proof, fenced-in yard, let your cats out on a long leash and harness. Better still, build an outdoor enclosure for your cats. This enclosure can connect with the house via a chicken-wire covered "cat walk" so the cats can go out and come in whenever they like. A flap door that is cut into a screen door or window frame with the cat walk connected to it is a very simple and effective design. Then, if the cats wish to spray, they will do it outside in their enclosures and offend no one.

Alternatively, a screened-in balcony or patio can give cats a little fresh air and freedom as well as a view of birds and other wildlife, which cats enjoy. Set up some carpeted shelves and a long tree branch for the cats to climb and recline on, or purchase a cat "condo."

Placing a resting pad and "cat condo" by a window provides much environmental enrichment for cats that love to look outdoors.
Photo M. W. Fox

*Excessive Grooming*

Excessive grooming is a habit some emotionally disturbed cats will develop, and they will lick themselves raw. This may be caused by some easily rectifiable change within the family, or it may be a side effect of spaying. Diazepam treatment prescribed by the veterinarian can help cure with a fear- or anxiety-based etiology. Prescribed hormone replacement therapy (progesterone) may sometimes resolve the latter problem. Many cats suffering from self-mutilation as a consequence of excessive grooming develop bald patches on their flanks. This can be a sign of a hyperactive thyroid gland, and in some instances a food allergy. Some cats will start to pull out tufts of fur when they are extremely upset; one cat I know did this every time one of her beloved human family companions went back to college! The ultimate cure was for the student to take the family cat with him.

Excessive scratching and rubbing of the ears, resulting in sores behind the ears, can mean a problem inside the ears, mites being the most likely culprit.

Obsessive-compulsive grooming can signal emotional distress.
Photo Aya Kinoshita

*Fur Balls and Vomiting*

During the course of regular grooming, cats swallow some of their fur and it is natural for them to later regurgitate small "sausage rolls" of hair, often accompanied by loud and disturbing sounds of gagging, retching, and coughing. This should be no cause for alarm. Many cats simply void the fur in their stools. Cats who frequently vomit up fur balls benefit from a daily grooming and a few drops of fish oil or a teaspoon of olive or safflower oil and a tablespoon of chopped wheat grass every day mixed into their food. Any cat who keeps trying to vomit without success, or refuses to eat, or vomits food soon after eating, should see a veterinarian at once. The cat may have a food allergy, or a excessive accumulation of swallowed fur in the stomach or blocking up the intestines.

*Sucking*

Like dogs, some cats will lick, chew, and swallow whatever they can to induce vomiting. But more often such behavior in cats, especially Siamese, is an obsessive-compulsive disorder, sucking and chewing on wool blankets and clothing especially. If the cat in question is not swallowing any significant amount of material, it is often best to give them their own blanket as a self-comforter to nurse on! More rarely cats will nurse on themselves, sucking on a paw or tail tip.

*Nymphomania*

Nymphomania may develop in female cats that have repeated heats, are not bred, and develop cystic ovaries. Life with such cats is difficult, to say the least, and the best remedy is spaying. It's a myth that cats (and dogs) need a litter in order to "improve" their temperaments. Symptoms include repeated rolling, licking the hind quarters, increased activity and irritability, loud vocal calling, and crying to go outdoors.

Un-spayed cats may go into a false pregnancy and adopt a toy or other object as a surrogate kitten, acting protectively toward the "fetish," and even producing milk. In such instances it is wise to remove the surrogate object and have the cat spayed to prevent such potential health problems as uterine infection (pyometra), cystic ovaries, and even ovarian cancer.

*Aggressive Behavior*

Jealousy can cause a cat to act aggressively, sulk, and avoid contact, or become unhouse-trained. Some cats do get jealous when another animal or person is getting all the attention. The jealous cat must be treated with patience and made to feel loved and secure again—exactly as one would with a child who exhibits sibling rivalry.

Aggression toward other cats in the home is a common problem in cats that are intolerant of other cats. Sometimes two cats who do not get along together improve when a third cat is added. But there is no way to predict such an outcome! Litter-mates and mothers and their offspring usually get along best. Cats tend to fight each other in the home more often when they are not related, and especially when they have free access to the outdoors and roam free.

Working out the social dynamics and situational triggers of aggression is the first step toward resolution. The cat pheromone product called Feliway, that can be administered as a spray or dispensed in a room diffuser, has been found to help cats settle down better together, reduce conflicts, and facilitate the introduction of a new cat into the home. Likewise, identifying the situation or context in which cats display aggression toward humans—such as when being petted, disturbed by a cat outside, or when playing and suddenly going wild—is the first step toward prevention. (Note: A cat's scratch can cause a medical condition called cat scratch fever, and a cat bite that penetrates the skin should be treated as a potential medical emergency, especially in people with impaired immune systems.)

Many cats seem to be acting aggressively, waiting in ambush to attack people's ankles. This is most likely playful prey-stalking and catching behavior, and is best dealt with by either regularly playing with the cat or getting the cat a cat-playmate rather than scolding the cat. Threats and punishment could actually incite the cat to really attack!

Sometimes one cat will mount another and seize the partner by the scruff or nape of the neck, especially during playful fighting. Such behavior should not be regarded as sexual (both male and female pairs will do it), but rather seen as a display of dominance and passive submission.

Older cats often show changes in temperament from easy-going and friendly to more irritable, aggressive, and restless, pacing and yowling especially at night. These are cardinal signs of what is commonly called senile dementia. This chronic degeneration of the brain in older cats, in ways sometimes identical to Alzheimer's disease in humans, is quite common. Helping the cat feel more secure by extra petting and comforting, providing warm sleeping spots, and having the veterinarian prescribe a psychotropic medication like Seligiline or diazepam can help this distressing condition that I call age-related dysphoria, also seen in old dogs.

Cats suffering from hyperthyroid disease, or chronic, painful conditions like arthritis, or acute, local pain, as from a bite abscess, can become suddenly aggressive when being petted or picked up. Any sudden change in behavior calls for a veterinary checkup, not discipline or physical punishment.

*Fear and Phobias*

Some cats suffer from fear of strangers or xenophobia, and this condition can seem to develop suddenly for no apparent reason. Young cats on the timid side who are not regularly exposed to strange humans from an early age, especially between ten and sixteen weeks of age, are very likely to run off in terror when an unfamiliar person enters the room. In many instances cats develop specific social phobias like being afraid of men or children, or other animals, because during their early formative months they were only around women or adults respectively. Since prevention is the best medicine, the best way to avoid cats from becoming socio-phobic is to expose them to as many different experiences and social situations early in life. Cats with specific phobias can be helped by qualified behavioral therapists and veterinarians using a variety of approaches ranging from appropriate anxiety-reducing medications to desensitization and behavior-modification therapy.

*Cradling Therapy and Training*

Simply cradling a kitten in one's arms is part of the process of animal socialization that is as gentle as it is profound. Cats and kittens soon learn to accept being picked up and gently held in one's arms without struggling, and they enjoy the intimacy and security of close physical contact.

Submitting to and accepting such handling is integral to effective and proper socialization or bonding with the human care-giver. It greatly facilitates subsequent training and communication. If and when the animal struggles while being cradled, the gentle embrace becomes firm resistance that immediately softens and yields as soon as the animal ceases to struggle, begins to relax and accepts cradling restraint, and then starts to trust.

This gentle psycho-physical "judo" can help in the behavior modification of adult, hyperactive, fearful, poorly socialized companion animals, often with a history of being overindulged and having no sense of boundaries and limited self-control (or what the Russian scientist Pavlov called *internal inhibition*). Cradling conditions the animal to accept restraint, develop internal inhibition or self-restraint, and, above all, develop the kind of trust that is the keystone for a strong, sustaining human-animal bond.

# 4
# | Animal Affection and Attachment |

## Imprinting and Dependence

There have been numerous studies on the development of emotional bonds (social attachments) in animals, and they give us a much closer understanding and empathetic appreciation for the social and emotional relationships between animals and human beings.

The phenomenon of *imprinting* is seen in most birds and mammals that are relatively mature when they are hatched or born. Because their ambulatory and sensory (smell, sight, and sound) abilities are so well developed, they respond immediately to their parent(s) as soon as they enter the world and become attached in a matter of hours. According to the species, the smell, visual configuration (body shape, colors, etc.), or sounds and calls made by the parent (or a combination of such signals) are the specific cues that bring about imprinting.

For a chicken or duckling, the movement, shape, and call of the mother evoke following and attachment. Once the attachment *imprint* is made, the offspring will ignore the calls of the other mothers and only respond to the call of its own mother. The same goes for the mother. Mammals that are mature at birth (for example, sheep, pigs, and goats) also rely on the smell of the mother. A caribou calf or a lamb can quickly find his mother among hundreds of others within hours of being born because the imprint of his mother's call has been established.

In most species, the mother is also imprinted onto the specific sound and/or smell of her offspring. For this reason, a mare or goat will refuse to nurse another's foal or kid, but she may accept an orphan if it is first rubbed down with a moist cloth that is tainted with her own body odor or of her offspring's odor. Similarly, shepherds will skin a dead lamb and tie it onto an orphan lamb so that the dead lamb's mother will not reject the orphan.

Imprinting is remarkably rapid and enduring. Sometimes, but not always, the attachment with the parent breaks at sexual maturity. Yet, ironically, the initial social imprint will often determine later social and sexual preferences. Mary's little lamb that followed her to school was imprinted onto Mary. Any presocial (mature-at-birth) animal raised by a human foster parent (or some other alien species) *before* she has become imprinted onto her own parent will become attached to her foster parent. A bottle-raised fawn, for example, will quickly become attached to people. This imprinting phenomenon is one of the major reasons for the difficulties (and hazards) of attempting to re-introduce a hand-raised orphan animal into the wild—be it a quail or a deer. The human attachment means that the imprinted animal regards human beings as kin, and the imprint may be virtually impossible to erase. In some species the attachment may break at full weaning or sexual maturity, but it is not unusual for the animal—even as an adult—to behave in a dependent, infantile, and solicitous way toward human foster parents.

Two other problems can arise in such human-attached animals. With maturity, their sexual behavior may be directed toward people—with predictable confusion, conflict, and frustration. Occasionally possessive (sexual) rivalry may occur, as when the human-attached animal is jealous of another human being who is close to its human parent-mate. Actual sibling rivalry can occur (with a human child or other young around), with status conflict and fights for social dominance. Here again, the human-imprinted animal will respond in similar ways toward people as it would toward its own kind in the same social contexts.

An imprinting-like, attachment-learning process in various animal species has also been observed in relation to a place or location (called *philopatry* in humans), and also to certain kinds of food. In birds, the same process occurs with certain sounds and complex songs.

*Socialization*

The social imprinting detailed here is analogous to a much more prolonged process of attachment that takes several days or weeks in other birds and mammals that are relatively immature at birth; starlings, eagles, rabbits, cats, dogs, and humans belong to this group. The attachment in these animals

Feral-born kittens like these may soon become wild and unapproachable if not given frequent human contact.
Photo M. W. Fox

is called *socialization*, and the same problems can arise between a human-socialized animal and people as those described for imprintable species.

The other most important practical aspect of imprinting and socialization is that there is a *critical period* for attachment, an optimal time when the bond with its own species and/or with humans can be established. In the cat, for example, this is from six to twelve weeks of age; if no human contact is given until a kitten is more than three months of age, she will not make a good pet since she will not "bond" well to a human companion and, consequently, lacking dependence, she will be difficult to train and handle. This is especially true for kittens born in the wild; having no human contact during these formative weeks, they can be extremely difficult to socialize with humans and may never establish a close bond.

An understanding of imprinting and socialization, which "cement" affection and allegiance in animals, including humans, can help us break down the species barriers and discover the richness of relationships—indeed a kinship—with animals other than ourselves.

## Touch and Love

Several years ago a noted pediatrician, Dr. Rene Spitz, became involved in the care of orphan infants. He found that in those orphanages where babies received little tender loving care but had all their survival needs attended to (clean diapers, baths, regular feedings, etc.), outbreaks of disease were common. Also, the babies didn't seem to thrive and some even developed a wasting disease known as *marasmus*.

Today, we have an analogous condition with many commercial animal breeding facilities. Young kittens that are separated from their mothers, even if given plenty of food and warm quarters, don't thrive as well and are more likely to succumb to diseases. Now we are beginning to understand this phenomenon. Part of the answer is related to the creature's heart.

When an animal is petted or groomed or licked by a companion, there is a dramatic decrease in heart rate. When you stroke a cat, provided she's not too excited, her heart rate slows down. This means that the parasympathetic side of the autonomic (vegetative or autonomic visceral nervous system) has been activated. Your touch can evoke such profound changes in the animal's physiology. These changes too must be pleasurable for a socialized cat (one who is not afraid of you), since the cat will approach and solicit such physical contact and social/tactile stimulation. (And we all need to get our strokes too!)

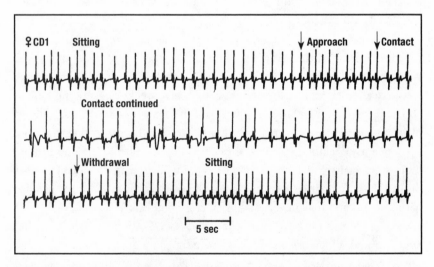

An animal's heart rate increases when approached but slows down dramatically when petted.
Foxfiles

When the parasympathetic system is stimulated, an infant animal relaxes, begins to secrete more digestive juices, and its alimentary system is activated to absorb food. Maternal deprivation, or lack of TLC, can therefore be detrimental to survival.

If food isn't properly assimilated in a young animal, susceptibility to disease increases. It would seem that these animals are born with two physiological dependencies that the mother normally rectifies by giving affection. Food and a warm bed simply aren't enough.

Dr. Spitz instituted a program of frequent cuddling for the orphaned crib-bound infants, and their health and growth rates immediately improved.

This physiological need for TLC also has another important consequence: attachment. The inborn physiological dependence upon the mother

Friendly adult cats lie together and groom each other, and an adult cat grooms an adopted kitten.
Photos Robin Scott

leads to attachment (since TLC is pleasant and rewarding), which in turn leads to an emotional or psychological dependence (upon the mother, foster parent, or caretaker).

Through this attachment process, imprinting or socialization takes place and an enduring bond is formed. This bond persists even in adult animals, and this is why socialized cats enjoy being groomed and petted.[1] It is through touch that humans and animals can appreciate and share a depth of nonverbal communication that transcends the species barrier, so they can reaffirm their kinship.

## Invisible Barriers

When you approach a wild cat or one who is terrified of strangers, he will be quite relaxed and unafraid at a certain distance. If you approach closer, he will take flight because you have crossed an invisible threshold and have penetrated his *flight distance*. If you were to approach even closer and the animal's flight is blocked, he may turn and attack you. This happens at close proximity when you enter the animal's critical distance or attack zone.

A bottle-raised wild animal, socialized to people, will not show these reactions. Socialization essentially eliminates the flight and *critical distance* reactions.

When this personal space is entered, there are certain laws or rituals to adhere to. As is the case with animals, people can only do certain things at close proximity; some parts of the body may be touched while others are taboo or are exclusive to very special or intimate relationships.

Cats during social investigation slowly approach each other to enter the close personal olfactory zone (to touch noses and sniff around facial scent glands). Such an intrusion on personal space increases the probability of conflict between cats who do not know each other or who are rivals. Displays of intention have therefore evolved, such as submissive crouching, a low and fearful yowl, displacement self-grooming or ground-sniffing, and making or avoiding eye contact to signal dominance status and intentions.

There are accounts of remarkable people who have the ability, as St. Francis of Assisi did, to approach wild animals and frightened cats and dogs

---

1. For details on the therapeutic benefits of touch in animals, see *The Healing Touch for Cats* by Michael W. Fox (New York: Newmarket Press, 2004).

while evoking no fear; animals may even come to such people spontaneously. What magnet attracts them, and what vibrations emanate from such rare individuals? Very often, as in the case of the circus lion tamer, the animal has been bottle-raised and is attached to and trusts the person. Consequently, the social and personal distances may be crossed without flight or attack. Many people who know cats stay still and let the cat approach them first. This may explain why people with cat phobias or allergies are approached by cats. The uneasy person sits completely still when in a home with a cat, and the cats almost invariably pick them out and approach, for being still is a cat signal of acceptance! When we raise a kitten from infancy, socialization creates the emotional bond, which acts like a key and enables us to make contact with them and, in turn, makes them approach us for affection. If deprived of such socializing human contact, a cat will react like a wild animal toward people.

Sometimes we can all be like St. Francis with animals, when animals have no fear of humans or of being hunted by other animals. On the Galapagos Islands, many of the resident fauna can be approached and touched by humans. Often, the animals of the Galapagos will come close to investigate human visitors. Explorers who discovered vast rookeries of seabirds and seals, wild canids like the Falkland Island wolf-fox, and the now-extinct dodo bird were amazed at the lack of fear these animals showed. Sadly, the early human visitors to the islands clubbed the curious little wolf-foxes to death, along with the dodos, and the seals were pounded to death—just as they are today, for their skins. Today's surviving seal colonies have learned to be wary of man—so they take to the sea while leaving behind their helpless calves, the skins of which are worn in ignorance of the bloody slaughter.

Many observers have noted the similarity between the flight distance of animals—how close one can approach before the animal flees—and the reactions of some violent, paranoid, and schizophrenic patients. Some of these patients feel that what is happening outside their flight distance is actually happening *inside their own bodies*. Their reactions include panic, escape, social withdrawal, and a need for solitude.

How wonderful it would be if all animals and people could come close in trust and love. Human beings have been hunters for thousands of years, and I believe this factor alone lead to a selection of those animals who were wary of humans. They would flee and therefore survive and bear offspring of similar temperament, while the curious and the friendly would be killed.

But today, only a small percentage of the world population still hunts; perhaps in a few hundred years more animals may come to trust us again. Such trust is seen today in the wild animals in sanctuaries in the United States and in areas of Africa, where hunting is prohibited. But don't get out of the jeep! The big cats are habituated to eco-safari tourists inside the jeeps, but one step out to pet them could be fatal.

## Animals as Mirrors

I am not sure by what process it occurs, but it is a fact that cats often "take after" their owners emotionally and temperamentally. This may be due to sheer coincidence, to careful choice, or to unconscious identification and selection by the owner. Developmental and socialization factors also play a part: namely, how an animal is raised and is thus affected by the owner's emotionality and temperament, and how emotionally dependent it becomes toward its human companion.

I have known companion animals that are emotionally affected by their owners' rage and depression, becoming fearful, aggressive, or withdrawn and depressed themselves. The more sympathetically attuned the animal is to the emotionality of its human companion, the more it can be harmed, or benefited, as the case may be.

I call this emotional phenomenon of mirroring *sympathetic resonance*, where the emotional state of one being affects another. This is easily demonstrated by a person scolding or playfully soliciting an animal who is emotionally attached to him—or to people in general. The closer or more symbiotic the attachment, the more vulnerable and responsive the animal will be.

Considering, therefore, that our emotional states can harm or benefit animals, more attention needs to be given to two aspects of the human-animal bond. One is the bond of stewardship (as within laboratory and farm animal husbandry). If there is no empathetic bond and the attitude toward the animals is negative (demeaning, controlling, detached, and objectifying) rather than positive (nurturing, compassionate, patient, playful, and understanding), then the animals may suffer or be harmed.

The other is the bond of companionship, as with a cat or dog. It may be emotionally exploitative—as when a person is overly controlling—and therefore harmful. Or the sympathetic resonance may be either beneficial (as with a playful and attentive owner) or harmful (as with a depressed, angry, paranoid, overdependent, hypochondriac, or otherwise emotionally disturbed and insecure person). The cat may feel rejected, depressed, afraid, and perpetually anxious and uncertain. Or the cat may become a fear-biter or develop psychosomatic disorders, such as asthma or chronic skin irritation, or possibly other diseases, ranging from heart disease to cancer to irritated or inflammatory bowel syndrome (with chronic diarrhea). The owner may also experience the same symptoms in the same organ systems. This correlation or coincidence of health problems in both the cat and the human companion should cause little surprise, since it is known that emotions and temperament affect the body (the physiology, metabolism, disease resistance, and organ/system diathesis or susceptibility). Genetic and other environmental factors, such as nutrition and allergies, also play a significant role in such problems, but they do not mitigate the significance of the emotions and of relations (human and animal, as the case may be) in the etiology of health and disease.

On a recent lecture tour of the Philippines, one veterinarian and anthropologist told me that some people in that country believe that an animal living with the family can take on, and thus take away, sickness in the family. This belief leads them to not take a sick family animal to be treated—for if it is cured, a family member would then be likely to become ill instead of the cat!

With these thoughts in mind, health care providers should perhaps take extra measures when "prescribing" animals for the emotionally handicapped and for an emotionally disturbed public at large, since animals can be harmed and can suffer. I do not suggest that such people should be prohibited from keeping companion animals—although some radical animal liberationists might hold to this ideal. Rather, I would suggest that in recognition of animal rights and the tenets of basic common sense and moral sensibility that all animals used for pet-facilitated therapy be under the care of a sensitized veterinarian—ideally in collaboration with a qualified animal psychologist/behavior therapist. Likewise, veterinarians in general practice need to appreciate the clinical significance of what I call sympathetic resonance.

Some animals may be more susceptible to the adverse effects of sympathetic resonance than others. It would be prudent, therefore, to develop objective criteria to screen out such animals that might be harmed if used in pet-facilitated therapy programs.

Our emotions, temperament, and attitude do affect the physiology, behavior, emotional state, and well-being of others—be they other humans or animals under our care. An important aspect of empathy is knowing how we affect others (through sympathetic resonance) and how others perceive us. This is, I believe, a long-neglected yet fundamental area of veterinary medicine and animal husbandry that warrants far more recognition and careful study.

In the next chapter I will give a more personal account of my experiences with one of the many animals who have enriched my life: Igor, my Siamese cat.

# 5
## | Lessons from My First Cat |

Now what is a freshly minted veterinarian to do for a housemate when newly arrived in a foreign country and feeling lonely and isolated in a small town? I worked all day with dogs at the Jackson Memorial Laboratory research station—baleful beagles, feisty fox terriers, soft and sensitive shelties, cringing cocker spaniels, and aloof African basenjis. So a dog was out. Anyway, a pup alone in the house would have been lonely and would have probably wrecked the place.

*House* is perhaps too pretentious a word to describe my living quarters. I had rented the "fully furnished" ground floor of a rickety two-story frame house just off the road at the edge of the woods in the outskirts of Bar Harbor, Maine. The furnishings were a spartan mixture of old motel pieces and pre–World War II hand-me-downs, the overall effect being that of a déjà-vu time warp that one can best experience in a Goodwill Industries showroom.

Bar Harbor was a ghost town in winter—all the restaurants being closed until summer, when the tourists would return and enliven the community. The shimmering gold and ruby maples, the azure skies, and the newness of my situation soon began to fade as the gray cold of winter invaded my spirits. The only place in town where there was any life at all was Butch's Bar, where Butch, who coincidentally was my landlord, served thin, fizzy beer in jugs and men talked of things that I could neither understand nor fake much interest in: the latest football game, the size of their lobsters, and the state of their clam beds and marriage beds. By mid-December, Butch decided to close up for lack of business, probably because his few patrons talked more than they drank.

So I felt even more alone in the New World without any bars or restaurants open. Such a contrast from Cambridge, England, where I had spent most of 1962 after graduation, as house surgeon at the new veterinary school. Apart from the intellectual and social stimulation at the research station, this place in winter was bereft of any kind of social life that was compatible with the tastes, needs, and aspirations of a young Englishman. My green-card designation of "resident alien" seemed ironically all too appropriate.

I found going to the movies alone almost as depressing as drinking alone, but since drinking was not my cup of tea (Butch's beer, no exception) I decided one night to break the monotony of winter and escape into the warm fantasy land of the celluloid screen-world.

*Bell Book and Candle* was showing at the local theater, starring Kim Novak and James Stewart. This charming film gave me hope and inspiration in finding a solution to my need for companionship, other than canine or human. It was not Kim Novak I fantasized about, for that could never have materialized into a satisfying reality, but the Siamese cat. I could surely buy a look-alike of the magical cat Pyewacket in the movie and all my problems would be over.

The first weekly local newspaper that I bought specifically for the purpose of its "Pets for Sale" advertisements gave me an immediate lead. That morning, in spite of snow warnings, I drove off in the 1956 Ford sedan Butch's uncle had sold me to a small community about an hour's drive away. By the time I reached the old Cape Cod–style home where a litter of Siamese cats was for sale, the sky was ominously leaden and a strong, cold wind was blowing. Mrs. Beal was more interested in my English accent and in telling me about her deceased husband who came from England than about my questions concerning the health and lineage of her cat and kittens. The kittens were certainly Siamese, as was their mother; but Mrs. Beal was quite uncertain as to the whereabouts, though not the identity, of the father. The mother cat purred at me, and louder still when I handled her offspring, as though she approved of my doing so—if not soliciting me to purchase one of her kittens.

I wanted to see the father to judge his temperament, since bad qualities, such as fearfulness, could be inherited by his offspring. The owner searched the house, and then led me to the barn. "Here Peter-Kin," she called repeatedly, and I imagined some refined seal point trotting toward us with tail up like a solicitous kitten. What I heard quickly shattered this image. A low, deep, penetrating, gravelly "meow," more like a roar of pride and presence than a mew of solicitous greeting, cut the dank, musky air of the cold barn. There, above my head, sitting like an Egyptian cat-god, unblinking, unmoving as though in direct communion with the ultimate mystery and truth of all creation, was the most incredible creature that I had ever seen outside of the wild.

"He's quite wild," said Mrs. Beal. "He does all his own hunting. Likes rabbits and brings them home, sometimes alive for me to kill. Or the kittens."

Peter-Kin roared at me, closing his shimmering blue eyes in complete acceptance of me: no fear, shyness, coyness, or distrust, but sheer presence and self-confidence. Then he sat up, stretched, leaned toward me, and touched my nose with the tip of his nose—a salutary feline greeting and a tradition for millennia. I raised my hand to pet him and he seized it firmly but gently in his jaws when I touched the base of his tail. No kittenish contact for him! He turned and rubbed his head and chin against my hand as soon as he released it, anointing me with the scent glands of his forehead, lips, and chin. This was another ancient feline ritual, a symbol of acceptance into his social sphere, for cats only mark their close companions and territory in this way.

At once, Mrs. Beal seemed to change visibly from a cliché-filled yet shy Downeaster to an animated and interesting human being. She was amazed at how Peter-Kin had immediately accepted me. Perhaps I reminded him of her husband, I offered.

But Mrs. Beal was adamant. Peter-Kin knew that I respected and understood animals, and anyway, her husband was long dead before she acquired Peter-Kin from a foreign sailor whose ship was being worked on at the Bath docks, where she worked as a secretary.

She told me that cats are psychic, that they have ESP and can tell if a person dislikes or likes animals or is afraid of them. "Real cats have all kinds of incredible abilities," she proclaimed, "least of all their ability to judge a person's character."

I felt flattered by the cat and inspired to have one of his offspring, but my scientific objectivity prevented me from accepting Mrs. Beal's pronouncements. She sensed my skepticism and observed, with sudden realization, that I must have never owned a cat before. If so, since I was so receptive to Peter-Kin and therefore he to me, I would have known all about the incredible abilities of cats.

Nonsense, I thought, as I suggested we return to the house so that I could select the kitten most like Peter-Kin. Yet was it all nonsense, since I had felt something incredible and ineffable in the presence of that sagacious Tibetan cat? The duality of self and other was transcended in an instant of relatedness as our noses had touched. He acknowledged me, greeted me, and then anointed me with his scent to make us one. He was a mystical, but also a practical, cat with the mind and physique capable of hunting throughout the Maine winter.

I squatted down beside the litter of kittens, all in a row at their designated teats, as their mother stared at me in open-eyed trust and then closed her eyes and began to purr. Carefully I observed the four kittens, noting which was the largest and most vigorous nurser. I picked each one up to see which was the most tense and frightened, or the most relaxed and responsive. The mother cat seemed to purr loudest when I held the one who was the most vigorous at the milk bar. This kitten was also the most relaxed and responsive.

"This is the one," I said confidently, and Mrs. Beal replied assuringly that she thought I was the right person for this member of the litter and therefore I could have the kitten. Had she thought otherwise, I knew that I could never have bought any of the kittens. "Twenty dollars, and take good care of it," she said.

I tucked the kitten up under my sweater, buttoned up my coat, and ran to the car, supporting the furry little heartbeat with one hand as I drove away into a blizzard of snow. The car had no heater. Suddenly I applied the brakes. I had not checked the kitten's sex, and I wanted a male just like Peter-Kin. How foolish, I thought. Some veterinarian I was, so preoccupied with getting a super-kitten as to be incapable of remembering to check the obvious!

I pulled the mewing kitten out from under my sweater, feeling its fine claws raking into my shirt to hold onto warmth and security. It mewed in my face, its breath making a little puff of steam in the chill interior of the car. Its eyes seemed to be steamed over like the side windows of the car and then I remembered: relaxed cats let the membrane in the inner corner of the eye extend over the cornea when they are contented or sick. Sick it wasn't, but the cold of the car quickly changed its mood. Its eyes cleared immediately and it mewed in shrill protest, scrabbling with its claws to climb along my arm back into my warm sweater. I twisted it around, lifted the tail and looked. And looked. I couldn't tell what sex it was. So much for five years at veterinary school, I thought. As I later learned, both male and female Siamese cats have the same dark marks on their rears, which makes it difficult to tell them apart. So I turned the kitten over and squeezed its stomach gently and saw two little swellings appear at its rear end. A tomcat he was, and Igor would be his name.

Just before I stuffed Igor back into my sweater, I saw something brown hop across his rump and onto my coat. Fleas, no please, and no thanks to Mrs. Beal! Fleas inside my sweater as well as a kitten would have been too much, so I nestled the kitten between my thighs and we drove on. The driving snow, swish, and scrape of the windshield wipers and Peter, Paul, and Mary singing about Puff the Magic Dragon on the crackling radio were so mesmerizing that I did not notice that Igor had left the warmth of my lap to explore the floor of the car.

Impulsive as usual, I had not made adequate preparation at home for a new cat: no litter tray, no litter, and no cat food. As I was contemplating further necessities a large truck carrying half-frozen chickens in wooden crates, fully exposed to the elements, from one of the many broiler chicken factory farms that dotted the Maine countryside came straight at me, plumb in the middle of the narrow road. I swerved and instinctively jammed on the brakes to avoid a horrendous collision, envisioning mangled chickens and smashed crates all over the road. The car hardly slowed down at all as it went into a gentle slide. I saw pine trees looming and then spinning as the car serenely careened in a semicircle around the chicken truck, leaving all unharmed and my car facing the opposite direction. Before I could collect my wits and curse the cretinous driver, I heard a half-throttled scream coming from under my foot. There was poor little Igor with his head stuck under the brake pedal. So that was why the car hadn't slowed when I put my foot on the brake peddle and felt something stop me from jamming my foot down hard. It was a wonder I hadn't broken his little neck. Wonder was, he may have saved my life, for had the brakes responded to my panicky over-reaction, I would have most likely gone right under the chicken truck.

It is ironic thinking of this incident thirty years and more later, since I am, at this time of writing, fighting for humane reforms in the livestock and poultry industries. And while not nearly so many chickens are now produced in Maine, they are still transported to slaughter too often under such distressing and stressful conditions.

The rest of the drive home proceeded very slowly, not because of the blizzard, which was abating, but because my mind was on my skin and the image of fleas crawling, sucking, defecating, and copulating under my clothes—for Igor was once more inside my sweater unharmed and safe.

I ran through the life cycle of the flea, recalling how the enormously equipped male flea copulates with a blood-engorged female, who, satiated and satisfied, lingers safely, usually behind the neck of her host where she cannot be reached and crushed between incisor teeth. She passes her eggs into the fur, along with her feces: dark flecks like coal dust that turn reddish brown and dissolve when combed out onto a sheet of wet white paper. I wanted to stop then and there and nip the fleas between my thumb nails: a tricky task, with eyes squinting, for engorged fleas burst far when squashed. Then I reflected on the fact that the eggs hatch into crawling larvae, microscopic maggots that live in cracks in the floor and in the pile of carpets, where they eat even more microscopic creatures and organic materials, including no doubt the myriad of minute particles of human and animal skin that collects in our houses.

Eventually the maggot becomes so fat that it cannot move: its outer skin hardens into a protective case within which the larva turns, not into some magnificent butterfly, but into an acrobatic survivalist and opportunist that will spend the rest of its life with free food, free love, warmth, and a free ride.

I was finishing reeling off the facts about fleas from the computer in my head, which thankfully still seemed to retain much of what had been programmed at veterinary school, when I suddenly stopped thinking (a rare event in itself) and began to feel fleas crawling along the insides of my thighs! Fleas biting, or rather my skin reacting to the effects of their saliva and beginning to itch.

Prudently, I stopped at the grocery store before going home to get what I should have bought earlier: cat food, cat litter, a litter box, plus flea powder. Flea powder they had, but in checking what it contained, I discovered that, even though the label said it was suitable for cats and dogs, it was actually poisonous to cats—and the smaller the cat was, the more toxic it could be. It could cause seizures, paralysis, brain and liver damage, and even respiratory arrest.

Half an hour later, Igor was on his back in my living room, pinned down by my knees and palms as my thumb nails searched and destroyed. His thin pale fur, made almost transparent by a reading lamp, made the fleas easy to catch: they stood out so clearly and the fur slowed their escape. One hopped off to escape, but quickly hopped back onto Igor, the apartment being so cold. Igor protested briefly and then went to sleep under the warmth of the light. I soon removed the fleas from Igor and quickly attended to myself, throwing my clothes out onto the back porch for the rest of the day. All this

The author and young Igor taking a nap, back in 1963.
Photo M. W. Fox

I knew was essential, since one flea could, if carrying eggs, quickly infest the apartment and have Igor and me as their domiciliary dinners, whenever they chose.

We never saw fleas again that winter, and Igor soon settled down once he had thoroughly explored the place and located the warm spots: the heat vents, the area under the desk lamp (when it was on), and the floor by various windows through which sunlight would create a pool of warmth at different times of the day.

As soon as I showed Igor the litter box, he sniffed around, pawed the litter gingerly, and then hopped into the box. He scuttled the litter under his hind feet, making a small pit as he squat, closing his eyes as he turned to face me. Suddenly his eyes opened and he turned around and carefully buried what he had just deposited, and with a quick shake of each paw, he skipped out of the box and came to me meowing, as though for approval or applause. I gave both, and following the rule of good petmanship, examined his litter box deposit. He was too young to have a problem of blood in the urine, a sign of bladder inflammation. But he was the right age for worms: roundworms and tapeworms.

Roundworms would appear in the stools like fine pointed strands and coils of thin spaghetti, while the segments of tapeworms would look like grains of crawling rice. I saw such mindless movement as I moved the litter aside with an old spoon. Igor had tapeworms and I guessed the kind. They were transmitted by infested fleas. He was too young to have a tapeworm from eating its cysts or mature eggs in the tissues of a field mouse or rabbit, which his father, Peter-Kin, might have brought home for his family. These malignant, motile segments, packed with eggs, would eventually dry and burst and be eaten, along with other microscopic things, by the maggots of fleas. They would remain as dormant cysts within the adult flea, their development suspended until the alimentary acids and enzymes of the cat activated it, after the cat had caught and eaten a "carrier" infested flea. Slowly, the scolex, as it is called, would turn inside out and its head, equipped with hooks and suckers, would burrow into the inner wall of the intestines. From this safe anchor would grow segment after segment of worm, tapes of segments interlacing and arising from a dozen imbedded heads. Like some sessile seaweed, the end segment of the worms would ripen and break off, to pass out in the stools, as younger segments took their place in a continuous cycle of regeneration, the surface of each segment absorbing the partially digested nutrients of the intestines.

What consciousness lay within those vegetating heads, I wondered. Little or none would be needed because the certainty of the relationship between flea and cat insured the worm's eternity. Yet, how could such a close inter-relationship between worm, flea, and cat ever evolve in the first place, I wondered, since such exquisite coadaptation must surely have taken longer than it took for cats to evolve into cats. But perhaps not. The life cycles of worms and fleas are shorter than the cats. There are so many generations of so many offspring that they can accumulate within the randomness of a diverse gene pool—various mutations and adaptive genetic changes that would insure their survival and continuation under the most adverse conditions. They quickly develop resistance to pesticides, I recalled. I thought of the lecture on parasitology by Professor "Scolex," as we called him, that I so rudely disrupted. Rabbit fleas, I recalled him saying in one rare, out-of-textbook aside, only engage in "love-making" when the female rabbit is producing hormones that signal she is ready to give birth and nurse a litter. This way, the fleas are sure of having more hosts for their offspring and have a neatly evolved birth control system.

Then there are other diseases that find their way from host to host through fleas, such as the plague, or black death, and typhus. Such dark thoughts, along with several feet of worms, were soon banished by appropriate medication that I picked up from the research laboratory the following day, since dogs, too, are afflicted by this same species of tapeworm.

Igor's great appetite and potbellied appearance did not change over the next week, however, so I again went through the ritual of raking the litter box and examining his stools. His coat was still thin and lackluster and, as with any first child, I was an overanxious parent. My paranoia paid off one morning when I found not tapeworm segments but a pack of shiny, curling roundworms in his stools. I recalled vividly the horror of finding worms writhing in the toilet bowel when I was a child. "Pinworms," the doctor had pronounced, and the cure was to drink volumes of nauseously bitter quassia chips (wood chips from some tropical tree). The impact on my young mind had indeed been traumatic: alien creatures living inside my body! And common though such worms are in kindergarten children, the whole idea was so embarrassing, especially to my mother. Perhaps it was this traumatic experience that gave me such a morbid fascination in parasitology, a subject in which I excelled at veterinary school and would surely have specialized in, had not Professor "Scolex" been such a crushing bore.

So Igor had roundworms, and I then remembered that kittens and puppies can actually be born with them, as well as pick up the mature eggs outdoors from contaminated soil.

Children can become infected by these eggs and suffer from a mild fever. They can even develop blindness after the eggs hatch out into motile larvae in the intestines and migrate into the body, finishing up sometimes in the eyes, presumably because they get lost in the wrong host. In pups and kittens they migrate to the lungs, are coughed up and swallowed, and then mature, eat, and breed in the intestines. Sometimes worms will become so numerous that they actually perforate the intestines and, of course, the kitten then dies an agonizing death. Somehow the larvae can cross the placenta and infest the developing kittens in the uterus. Such had been Igor's fate. After giving him a pill that evening, he evacuated so many worms that I could not believe his small frame could support such an infestation for so long without bursting or collapsing.

Within hours he was more playful and responsive. By the end of a week, his appetite was down, his coat was blooming, and the potbelly gone.

I then vaccinated him against common cat virus diseases (cat influenza and feline distemper) and learned that you don't inject Siamese cats on the back because the fur at the site of the injection will turn dark brown.

I should have vaccinated him on the underside or thigh where the fur is dark. And so he was to wear the mark of my incompetence for the rest of his life. It was no solace that few people ever noticed the tiny blemish on his coat: I always did.

As he matured, his coat turned darker, slowly replacing his pale creamy kitten coat, spreading from his tail, ears, and paws—a genetically-linked phenomenon related to temperature sensitivity where the cooler extremities acquire much melanin pigment early in life. This gradually spreads until the cat is dark all over, or stops halfway, leaving a handsome seal point pattern.

When Igor was twelve weeks old, I made a painful decision. He had to be neutered, since I knew that if he were not "fixed," he would want to roam and would spray the apartment with his urine as a territorial mark. Magnificent though I thought he would eventually be, I had no intentions of breeding Siamese cats and felt that the sooner I "fixed" him, the better. If he never had sex, he would surely not feel deprived. And so with mixed feelings of resignation, determination, and guilt, I quickly operated on him one Sunday morning, as sun and church bells filled the living room, bringing back memories of Sunday mornings in England when, from my early teens, I would accompany Mr. Howe, the local veterinarian, or one of his associates, on house calls and visits to outlying farms. Igor's recovery was uneventful, and by the following evening he was as playful and crazy as ever.

While a Freudian analyst might find theoretical affirmation in my preoccupation with the litter box, a veterinarian is by training impelled to concentrate on the basic realities of life, especially when dealing with animals that have a more limited capacity to express signs of sickness or distress. I was distressed for several weeks when Igor developed diarrhea each evening. I was concerned that this might be toxoplasmosis, an intestinal infection that can be transmitted to pregnant women and cause birth defects in the developing fetus. But a much more common source of infection is from contaminated meat. Igor's fecal samples all checked out negative, so I suspected a psychosomatic problem.

Every evening after dinner Igor would have the "evening crazies," which I later learned were characteristic of many felines. Perhaps it was some atavistic trait related to the nocturnal hunting behavior of Igor's ancestral lineage. He would race through the apartment, ricochetting off the walls, leaping at spots and shadows, and stalking and pouncing on imaginary things in front of him. A wall mirror was used to intensify his excitement. Slowly approaching it with arched back, he would fluff up all over as soon as his reflection came into view and then race away as though terrified, only to repeat the game a few minutes later.

Igor's behavior, which so excited him as to actually cause acute indigestion, was a clear indication to me that animals are not only imaginative but also highly creative in their play. I solved the diarrhea problem simply by feeding him after he had acted out his evening crazies. And I decided to join him in the game. As first he was unsure, seeing me on all fours poking my head around the corner of the sofa and bobbing up and down, as though to say, "Now you see me, now you don't." He skittered away with his tail arched in an inverted U into the bedroom. I followed and just as I reached the door he leaped on my back, batted my head with his front paws, and was off, racing around the living room with his tail all fluffed out. I waited, hiding behind the door, and he waited behind the sofa. It was a long standoff until he emerged, crab-walking like a Halloween cat toward me, looking as ferocious as when he stalked his own reflection in the mirror. At first I was disconcerted. He seemed to have become totally wild. I spoke to him in a calm, reassuring voice and he immediately became a little kitten again, switching from Hyde to Jekyll at the sound of my voice. I then hissed, raising my arms menacingly as I stood up and approached. He stood his ground until I was just one stride away and then he leaped into my arms, purring and biting my hands gently. I grabbed the fur and skin of his tummy and he held onto my arm with his claws, kicking with his hind feet as though tearing into some imaginary prey. Yet he had complete control of fang and claw, leaving no mark as we wrestled together.

Every night from then on we would play hide-and-seek, stalking each other and then finally closing in for the kill, after which we would appropriately have our dinner. This evening ritual was always disconcerting to visitors who came by for dinner. Most were patronizing of my eccentricity: few seemed to realize that to live with a cat, one must perforce, at times, become a cat too.

Playing together established a deep bond of affection and inter-dependence. Igor became so attuned to my habits that by springtime I needed no alarm clock to wake me up. His cold nose would touch mine and I would awaken instantly, almost precisely at eight o'clock. He learned the sound of my car and was always at the living room window as I drove up, and then at the door to greet me when I came in. He would object when I gave more attention to books and papers than to him, by jumping on my lap and sitting on top of my work. He taught me to throw balls of scrap notes that I would crumple up and throw on the floor, by retrieving them right to my feet. He would patiently play this game as long as I worked—and if I didn't, he would be up on my lap to interrupt my writing or reading without hesitation.

One of his favorite games was to put one of his front paws into my mouth while he was being held and nuzzled, for me to bite. He would then bite onto my nose and hold on, increasing the pressure only when I bit a little harder on his paw. For me, this was the most intimate lesson he gave me to demonstrate his trust, sensitivity, and great sense of humor.

As soon as the woods at the back of the house were clear of snow, I would take Igor on long walks. He enjoyed these outings immensely, and he soon knew what "go for a walk" meant—for as soon as I said the words, he would be by the door meowing to go out. I would never let him out free and unsupervised. He could get hit by a car, caught in a trap, or get injured in a fight with a dog, raccoon, a giant Maine coon cat, or even a bobcat. And I didn't feel it right for house cats to hunt and kill wildlife.

Igor would follow me just like a dog, or run ahead and explore and then wait for me, either in ambush to attack my legs or to be carried over my shoulders for a while. He started to talk to me then. I would say something like, "It's a beautiful day," or "Smell those pine trees," and he would meow back right into my ear several times in a low, penetrating tone. Siamese cats irritate many people because they are so vocal, but it is quite obvious that they simply want to converse. It's not so much what one actually says but how one says it that seems to be important to such cats. We would also go on long drives together, with Igor perched on my shoulder or on the rear window ledge, talking most of the time.

I came closer to understanding why cats purr when I was confined to bed with a severe attack of influenza that spring. I was worried that Igor might become infected also. I recalled that some flu-like viruses can afflict both cats and people, and that in veterinary practice in England we would often find ourselves doing many house calls to treat sick cats whose owners were also showing similar symptoms. Igor did not succumb but was a most attentive companion. He seemed to know that I was sick and spent nearly all the time lying close beside me on the bed, getting up occasionally with a gentle "meow" to touch my nose, rub his head against my chin, and give me a few licks with his raspy tongue. Then he would purr deeply and so resonantly that it seemed as though I was lying on a vibrating bed. The sound waves filled my body and I would immediately begin to relax. The experience was like being groomed or massaged by sound, and I realized then that cats no doubt evolved purring as a way of maintaining intimate contact without having to touch or groom each other. I even wondered if Igor's frequent bouts of attentive licking and purring were the cat's natural way of healing through love. Sick cats will fast and rest as though in a deep trance, and companion cats will often behave toward such sick companions just as Igor did with me.

Later I did some research on this and discovered that when an animal is talked to or petted, its heart rate decreases dramatically. Such a dramatic change in physiology has a beneficial healing effect. Tender loving care can heal. It is part of the miracle of the laying on of hands, and purring and licking are feline equivalents of the healer's touch.

Holistic medicine is today discovering that not only touch can heal, but also certain sounds and colors. In other words, subtle energies can help repair and harmonize mind and body.

In my febrile state I imagined Igor as a Tibetan monk. His purring actually reminded me of Tibetan monks chanting mantras and "chording" to produce the feline equivalent of a purring sound from their vocal chords. Perhaps cats induce or express some altered or transcendent state of consciousness when they purr. It is their way of chanting their mantra. Either way, it felt good to have Igor attending to me with such unconditional affection when I was sick.

It was at that time that I learned more of the vocal repertoire of cats. Igor would give a short chirp-meow to announce he was coming just before he leaped onto the bed. And he would chatter in bird-like cheeps and twitters whenever birds perched by the bedroom window. I still don't know to this day if cats try to mimic birds in order to get close to them for the killing pounce. Igor could bark too, I discovered while I had the flu. He gave a snuffly kind of explosive grunt as though to warn me when a friend knocked on the door to drop by and see how I was recuperating.

These and other experiences of living so close with such an empathic and playful creature as Igor were a revelation to me. It cast a bright light on my research on the behavior and brain development of the dog. I realized that to really study any living creature, one must combine scientific objectivity with respect and love for the intrinsic nature and worth of one's research animals. The more I knew Igor as an animal, as an intelligent, sensitive being, the more concerned I became about my own research and my scientific colleagues' way of treating their laboratory animals like unfeeling machines, mere "tools" for their investigations. I could not accept that my close relationship with Igor was some artificial or exceptional circumstance that somehow changed the cat. Rather, I felt that Igor, through the close bond that we established, had actually changed me. He revealed something to me of the nature of all animals when we earn their trust and affection.

I became increasingly uneasy about certain experimental procedures that caused unnecessary suffering. And the social and emotional privation of research animals kept in small cages for years, which I witnessed in other laboratories that I visited, began to burn in my conscience. Fortunately there were no cats in my laboratory and the dogs were kept under superb conditions in large pens and outdoor runs. I also began to question my own values, thanks to Igor, and soon came to the conclusion that it was ethically unacceptable to subject animals to any form of physical or psychological suffering or privation simple to satisfy scientific curiosity and to acquire knowledge for knowledge's sake. I was familiar with all the rationalizations: that knowledge is a cultural value; that tests on animals for new consumer products help protect consumers; and that scientific and medical breakthroughs can only come from research on animals, but not if scientific "freedom" is limited by humanitarians placing legal constraints upon animal experimentation. I was in a bind myself, realizing that my professional advancement (I was working toward a PhD in medicine as an external student from London University) was dependent upon animal exploitation. Thanks to Igor, I resolved at that time to avoid subjecting animals to suffering by finding alternative ways to acquire vital knowledge essential to the goal of advancing veterinary and human medicine.

It was only a few years later, after further experiences with cats, dogs, wolves, and other animals, that I refined this personal ethic to the point that I would never do anything to an animal, in the name of science, that I would not subject myself to voluntarily—even if there was *convincing evidence* that the knowledge so gained would be of benefit in alleviating sickness and preventing suffering in the world. This may sound overly idealistic and altruistic, but when you think of it, most human sickness and suffering is man-made. With appropriate changes in our lifestyles, values, dietary and other consumer habits, and in our technology and industries (which create so much ecological chaos, pollution, and sickness), we wouldn't need to use animals in biomedical research.

At last the late Maine spring turned quickly into summer, and the town came out of hibernation—with restaurants, bars, and tourist shops opening one by one fully restocked for the tourist invasion. For me, the long cloistered winter was over, and Igor and I enjoyed life to the fullest. We made many new friends and he approved of one young woman in particular, whom I eventually married. But the day before the wedding, a houseguest inadvertently left the kitchen door open. Igor ran out, was hit by a car, and killed instantly. Strangely, he died at exactly the same spot where, a few months earlier, he had slipped out and been run over. After that first incident, my fiancée Bonnie Morrill and I nursed him out of a semi-coma day and night for over a week, and he recovered.

But after the incident that killed him, we buried him in the Maine woods and it felt like a major chapter in my life had closed. Igor had been one of my best friends and teachers, and I resolved that someday I would find a way to express my gratitude to him and all his kind.

# 6
## | How Animals Mourn and Express Grief |

There is a beautiful bronze statue in the city of Edinburgh, Scotland, in memory of a little Cairn terrier named Bobby and erected by loving citizens who knew the dog. Bobby was revered by all who knew of him, for after his beloved master died in 1853 and was buried in the small graveyard of Greyfriars Kirkyard, Bobby took up his lonely vigil, lying beside the man's grave until he died of old age in 1872. His vigil of devoted loyalty lasted some nineteen years, according to the brass plaque on the memorial statue. (A commemorative headstone to Bobby in the graveyard puts his age at sixteen years when he died on January 14, 1872—but that discrepancy is academic, perhaps reflective of the Celtic calendar!)

Outside of a subway station in Tokyo, Japan, I was shown a statue commemorating the vigil of another dog people knew well because he would wait at this station every evening for his master to come home from work. For several years after the master died, this dog kept up his patient vigil, waiting for his beloved human companion to accompany him home.

The fact that people put up monuments to dogs in these far corners of the world says something about us, something that is good: we recognize, across cultures and time, the good that animals can embody, namely those human virtues of faithfulness, devotion, unconditional love, and selflessness. These qualities are seen by some as the domain of saintliness in our own kind, since these qualities seem all too rare in contemporary society. But perhaps they are not rare, as long as we continue to let animals help us become more fully human, their well-being under our domination (or dominion, biblically speaking) being a mirror of our evolution and development as a compassionate, humane species. As the Australian aborigines say, "Dingo [dog] makes us human."

Archaeologists have found, in Stone Age burial sites from Denmark to Cyprus, the remains of humans curled around those of an evidently beloved dog or cat. So clearly, from before the recorded, written history of humankind, companion animals were important to our kind and were venerated for their gifts of soul and spirit, presence and prescience.

The following selections from the scores of letters and personal stories I have received over the years spell out in graphic and moving detail how many, but not all, companion animals react to the loss of a human or animal companion and loved one. There is nothing sloppily sentimental in these testimonies, nor is there evidence of anthropomorphized projections of people's grief onto their animals. The scientist and secular materialist could reject these accounts as being subjective and unverifiable anecdotes and cast them into the fires of rational objectivity. But the animals, in their behavior, clearly speak for themselves in the ancient and authentic language of the heart that we share with them. Being true to their natures, they cannot lie.

Anyone feeling discomfort about the emotional depth and intelligence of animals will be moved to bear witness and to accept that we humans are not the only ensouled, sensitive, and intelligent beings on this little blue and green planet. The animals, in their grief, and in their longing for the return of the loved one, show us that the nature of love is an inborn quality that we share with them. We are moved to a deeper understanding and respect for them, since this means that they too must also suffer loss and experience joy and camaraderie in ways quite similar, if not identical, to us. Outward appearances can indeed be deceiving.

The loss of a loved one, human or nonhuman, is usually first expressed as acute grief, the animal making sounds of evident distress. Some animals often show no initial reaction but may later start to search everywhere and become more vigilant, apprehensive, and pensive—sitting and waiting by a door or window as though expecting the loved one to return.

"My cat Speckles," writes Nancy Falzone-Cardinal of Springfield, Massachusetts, "was six years old when our ten-year-old terrier, Whisky, died. They had been very close and friendly with each other. When our darling Whisky died, Speckles spent every day of weeks searching our home. She wandered from room to room, looking around and over everything, all the time meowing a dreadful cry. My husband and I decided to remove our living room recliner from our home because Whisky had spent his last month lounging there, and we thought his aroma might be torturing the cat into looking for him. After doing so, she stopped the entire apartment search, but Speckles then went to the front door every late afternoon, pacing and lounging within two feet of it—and meowing until 10 p.m.! For four entire months she waited for her 'brother' to come up the stairs from a very, very long walk. One day she just stopped her ritual. She lived another eight years. She tolerated our shih tzu, Randi, who joined our family, but she truly never loved him, as she had Whisky."

Kirstin L. Wesfall in Washington, D.C., wrote: "I had purchased two Blue Point Siamese brothers, Sparta and Troy. From a kitten, Troy had battled liver disease for several years and was the first to die. Upon his death, Sparta would go to the basement and let out gut wrenching screams. He would do this several times a day for an entire month and it just broke my heart. I have never heard a cry like it; I quickly realized he was mourning his brother. I have always believed pets grieve and I was certainly witness to that."

D.K., from Hollywood, Maryland, had two German shepherd dogs, Princess and Shep. "Princess was a year older than Shep and was the first to go. Shep was devastated; at times during the day he would go into the dark garage and moan the saddest moans that you can imagine. We got another adult female dog right away, but that didn't help. It took a few months for him to get over his grief."

Sometimes animals seem to know when another is going to pass on and will be especially attentive, like the account of Kisses, who comforted Orville in his last days. "I had a yellow Lab, Orville, who was quite old, and growing more feeble every day. I had adopted a one-and-a-half-year-old chocolate Lab, Kisses, only weeks before Orville died, but they barely knew one another due to Orville's advanced age. Kisses had a soft, stuffed goose which she carried in her mouth constantly; she slept with her head on it every night, it was obviously her comfort/security thing.

"One evening Kisses took her precious goose and placed it at Orville's head. For the remaining three days of Orville's life, Kisses would come by and sniff the goose and push it toward Orville, but she never made any move to take it from her new friend. The goose was at Orville's head when he went to sleep for the last time. Kisses took back her goose only after Orville was buried."

Sometimes the grief may be so intense that the suffering animal behaves in a totally disoriented and even reckless manner, like the cat BooBoo who

BooBoo (right) was so grief-stricken over the premature death of his companion (left) that he chewed off his own tail.
Photo Sabrina Campbell

belonged to Sabrina Campbell of Alexandria, Virginia, who bit off his tail when the older cat in the family died. Jean W. Burdell's cat in McLean, Virginia, became aggressive toward her, "with many ambushes and frequent puncture wounds," immediately following the sudden death of her husband. "Ever since then, whenever guests come to the house, the cat roams around excitedly, reaching up to the men present to attract their attention. His attitude toward me has progressed slowly to tolerance, acceptance, and even affection. Today I saw him again, as I have seen him many times, sitting in the closet beneath some of my husband's clothes looking up at them. I believe that in his primitive fashion he remembers, he waits, and he hopes," she writes. I would say, not so primitive—and get rid of the clothes!

I am reminded of my days studying the behavior of wild canids—foxes, coyotes, and wolves—and the morning I found one beautiful little kit fox dead in her enclosure. Her mate had piled food and various toys around her lifeless form and threatened me when I entered the enclosure, clearly being protective of her.

This caused me to reflect on the suffering that hunters, trappers, and others who kill wild animals bring upon the animals who are not killed. How many grieve the death of a mate, parent, pack leader, or beloved offspring?

Wildlife managers have repeatedly witnessed the grief of elephants (like us and some dogs, they actually cry when they are stricken with unbearable emotional pain and anguish) after herd-mates have been shot as part of the wildlife conservation management policy in East Africa, according to the culling protocol of population control. Wildlife managers now kill entire herds so there will be no survivors left to suffer. Alternatively, the whole herd—a nexus of close family ties amongst mothers, daughters, fathers, sons, grandparents, uncles, aunts, and cousins—is moved to another area to repopulate a depleted area. Population control is a management dilemma in many wildlife parks, elephant ivory poachers not withstanding.

Elephants have been seen protecting injured and dying herd-mates, burying the dead under dirt and brush, and gently touching the bones of the deceased when they come across them in their ever-shrinking domain.

Sometimes the grief that animals experience over the death of a loved one is so intense that they die from what we call, in our kind, a broken heart. D.R. from Moorhead, Minnesota, wrote to me about the two horses, Pete and Florie, that she and her husband kept to plow their fields. "They were well up in their years and 'retired.' One morning as my husband was doing chores, he noticed that Florie was staggering by the barn and promptly toppled over and died. Pete walked over to her and paused for some time then took a walk through the pastures and returned to Florie. He paused again and then put his front foot upon her body and fell over beside her and died too. That showed me how sensitive animals really are and that in their own ways, they do grieve also. As I'm writing this it still brings tears to my eyes."

June Baker in Arlington, Texas, witnessed a similar event with her father's dog: "Family members would take turns staying at the hospital with my father. Every time we returned home, Dad's little brown dog [a mixed breed] would meet us and smell every inch as though he knew we had been with Dad. When we all went to the funeral and returned home, his little brown dog came to every one of us and sniffed at each of us. Finally, he asked to be let out of doors. When he did not come back to the door to ask to get in, someone went out to look for him and found him dead under a bush outside. There were no marks or injuries and he was known to be in perfect health. We consulted a vet about this strange happening and he told us that sometimes, when a master dies, the pet will suffer a brain hemorrhage due to grief."

How this dog died was probably due to what is termed *vagal syncope*, the vagus nerve being part of the autonomic (subconscious) nervous system that slows the heart down and can even stop it. This is what often happens when people faint from shock; the heart rate slowing or momentarily stopping and the blood pressure plummeting results in a blackout (the opposite of a brain hemorrhage). So Dad's little brown dog died from cardiac arrest—a broken heart indeed.

Dorathea Rubsky from Miami, Florida, wrote a letter to me about the collie dog in her family that moped around the house for weeks when her father was hospitalized. "At first, we thought she might have been sick, but upon being examined by the veterinarian nothing could be found to cause the problem. Needless to say the night before my father passed away our collie died. Thus we figured she was mourning for him."

Other species can suffer, too, from the death of a loved one, showing clear evidence of depression. Doris Addenbrook's male canary in Virginia Beach, Virginia, fell silent after Mr. Addenbrook passed on. The widow says the bird, who has not sung now for eighteen months, "will peep if a male comes into the house or if I put him on the porch on a sunny day. I have played canary song recordings for him and I try talking to him. It is so sad to have lost his songs."

My friend from Adirondack, New York, psychologist Dr. Emmanuel Bernstein, tells me that he knew a goldfish named Gertrude who was used to Pierre, the parakeet, visiting her every day and pecking on the bowl. Gertrude would wiggle her tail in response and stayed opposite the bird wherever Pierre moved around the bowl. When Pierre died, Gertrude visibly mourned, facing away from the side of the bowl where there was movement for many weeks before relating to anyone else.

## The Wake and Viewing the Body

Some years ago I met a neighbor walking her three dogs. She was crying. Her fourth dog had just died in her home, and she was not only grieving the loss of her old dog but was worried about how the others would take it—for they had not yet seen the deceased animal in her upstairs room. I advised her to let the dogs in from the yard when she got home and let them see the body because they needed closure. If they never saw the dead body, they might wonder where the old dog had gone and be anxiously awaiting his return home. I saw her a few days later and she told me, "Doctor, it was just like a wake, when people come to view the dead. I called the dogs into the room and they walked slowly up to their dead companion I had lying on a blanket, sniffed him all over, then slowly and quietly walked away. I think that really helped them not miss him so much and understand what had happened to him."

Norma Nelson of Tenafly, New Jersey, writes, "After one of my cats died I left him in his burial box for our other cat to see. He stayed by the box for twenty minutes, then slowly walked away and was fine . . . Animals have to mourn their friends in the 'casket' before burial just like people do."

In Bloomfield Hills, Michigan, Janise Stoneman's bichon frise stopped eating and searched the house looking everywhere for their rabbit, with whom the dog had become friends, after the rabbit had died and was put in the garage because the earth was too frozen to dig a grave. After a friend suggested her dog would be less distressed if allowed to see the dead rabbit, she wrote: "We took our dog into the garage and showed her the rabbit. She stood totally still, not even moving her tail. She did not make a sound. She once turned to look at me with a sad expression on her face and immediately turned her attention back to the rabbit. After about ten minutes, she came back into the house. For the next couple of days, by habit, she would look to the empty cage and, as if reminding herself the rabbit was dead, would walk away. She no longer searched for the rabbit and resumed eating."

Jean Styles of Punta Gorda, Florida, sent me the following touching story about her terminally ill husband who decided to die at home. "Our miniature poodle, Mitzi (now deceased), loved us both but had an especially close bond with my husband. During the final days I would zip home at lunch hour to see if my husband needed anything and I would always find Mitzi parked right beside his chair like a good little nurse. When I would get home after work each afternoon it was the same story. Mitzi sitting close where she could keep careful watch over him.

"One day, while he could still communicate, he told me he didn't want Mitzi to die of grief when he was gone. To prevent this he made me promise that, when the time did come, before calling to report his death I would put Mitzi up beside him and let her check out the situation for herself. He was a very wise man but I still didn't see how this could help . . . On the fatal morning, after making sure he was no longer breathing and before calling for help, I put Mitzi up beside him. She sat looking at him for several minutes and then slowly got down and lay on the floor beside the bed.

"Mitzi had always known just when one of us was expected home and waited, perched on the back of the couch by the front window, for his car to pull in. I was afraid she would wait for him still but, to my amazement, she seemed to completely understand and never again sat there waiting for him to come home, although she still kept to that practice till I got home each day.

"I can't help feeling that she is now with him and the two are waiting for me to join them.

What a reunion that will be!"

Jean Milnor of Minneapolis, Minnesota, has a similar story. Her husband was dying at home from cancer and their dog, Hinny, insisted on getting up on the bed with her husband a couple of hours before he died. She writes, "Hinny snuggled up by his shoulder, she licked his neck, the hollow of his cheek, and then settled down to snuggle right next to him. She hadn't done this before. She laid in this spot until my husband passed over to heaven. After his last breath, she got up and moved down to his feet to lie there. There is an emotional connection—perhaps spiritual—between our dog and our spirits. I am honored to have witnessed the sacredness of the experience."

A story about a bull named Barnaby made the international news in 2004. This bull left his field in the German town of Roedental and found his way somehow to the cemetery, over a mile away, where his farmer-owner, Alfred Gruenmeyer, had just been buried.

Apparently the eccentric farmer treated his animals like pets, allowing them to have free run of his home. The bull had to jump a wall to get into the cemetery and insisted on staying there for two days in spite of the best efforts to get him to move out.

Carol A. Ross of Lincoln Park, Michigan, writes that some time after her father had died and was buried, she took his dog Rusty to the cemetery, "and like a rabbit, the dog ran directly to my Dad's gravesite. Remarkable or just plain dog sense?"

There can be no doubt that animals possess some understanding of death, and from my own personal experience I can vouch for the importance of having animals see their loved one, human or nonhuman, when laid to rest. But they should not witness the burial. I made that mistake with my dog Tanza, who saw me burying Quincy, my brother-in-law David's Seeing Eye dog whom my wife Deanna and I had taken in for a year after he became too old to help David. Tanza as a puppy looked up to him as her patient playmate and comforter. But when she saw me putting him into the grave, she screamed and frantically tried to dig and pull him out of the hole. Needless to say, I was devastated by my own insensitivity, taking the notion of closure too far. Tanza had been lying beside Quincy's body for half an hour after I euthanized him (and she and the other dogs did not witness that). I should have simply lead her away before returning dear Quincy to the earth.

Animals not only grieve the loss of a loved one. Some clearly know when the loved one just died; like the dog Hinny and the cat Boo from Schenectady, New York, they are able to anticipate death. Jaqueline Rosenbaum's husband was dying from lung cancer; she writes that, "A couple of days before my husband passed away, Boo (an eight-year-old Scottish Fold) was sitting on an end table at the head of his bed and howled the most horrible cry I ever heard (like he was in extreme pain). I know at that time my husband's system was shutting down and I'm sure my cat sensed it. (He has been depressed ever since my husband died.) I recently bought a kitten to help Boo (and me) through our grieving. Occasionally Boo will play with him, so I hope things will improve."

Boo's prescience—the ability to foresee impending death—leads us into the deep heart of animals who, as the next chapter documents, are able to somehow know that a loved one has just died—even though they are not present at the time of death. So does this mean that animals have psychic powers? Make up your own mind after reading the next chapter.

# 7
# | "Psychic" Animals and Their Super-Senses |

The more we learn about the mysteries of the animal kingdom that science explores and about the mysteries that animals reveal to us in their behavior, the more we stand in awe of the inherent wisdom and complexity of manifest life. Some of the abilities that animals reveal still baffle science and plunge us into the realm of the inexplicable, even the spiritual. *Psychic trailing*, or *psi-trailing*, is one of these.

It is quite distinct from *homing*, a superficially similar phenomenon that was for a long time thought to be "psychic." Homing is an animal's ability to find her way home after she has been deliberately released (like homing messenger and racing pigeons) or lost (like a cat or dog that slips out of the car on vacation with the family). Scientists have demystified homing by showing that various animals can use the sun, moon, stars, and the Earth's geomagnetic field as clocks and compasses.

Cats and other animals are sensitive to electromagnetic and geomagnetic fields, and they possess an internal directional compass and an internal time clock that enables them to have a sense of time in relation to where they are and the position of the sun.

There are iron (ferrous) salt deposits in the frontal brain region of birds, cats, humans, and other animals that act like a magnetic compass and thus provide an inborn sense of direction. This sense may be architecturally and technologically disrupted by, for example, people living and working indoors for extended periods surrounded by an electrical field and metal structures, and possibly by people sleeping in an east-west direction rather than a north-south one. Animals also possess an internal circadian time clock that enables them to fine-tune the position of sun (time of day) with the internal setting of their circadian clock. This internal clock is integrated, I believe, with the compass-sense to give animals an awareness of their place in time-space.

This time-space information is actually communicated by some animals to each other. For example, honey bees with their elaborate dances tell fellow workers how far away a particular food source is and in what direction. Scientists first found that bees are aware of the position of the sun and then, that they are also sensitive to gravity. Bees, like birds and humans, have iron deposits in their nervous systems.

## Incredible Journeys

Cats are especially adept at traveling long distances to find their way home after getting lost on a trip with their human companions. More often, not liking their new home, they set off and eventually reach the old home. They may be more motivated to do this than dogs, because dogs are happier attached to the human family-pack while cats become especially attached to their familiar surroundings and territory.

The longest journey that I know of was reported in June 1985. Distinctively marked with splotches of white hair among his black fur that led him to be called Muddy Water White, this cat was about a year old when he jumped out of the van being driven by the owner's stepson near Dayton, Ohio. Almost three years to the day, a bedraggled cat appeared on the doorstep of Barbara Paule in Dauphin, Pennsylvania. Even though he plopped down like he owned the place, she took him to be a stray until he was cleaned up. Her veterinarian, who had seen the cat back when the cat used to live there, was confident that this was indeed Muddy. Somehow he made it home over a distance of 450 miles.

Jasmynne traveled 300 miles to get to her old home in Ashland, Ohio, even crossing a bridge over the Ohio River after she ran away from her new home in Louisville, Kentucky, according to a newspaper report in June 1998. Over a month after her disappearance, her owners, Russ and Debbie Brown, were in slow traffic returning from Ohio after visiting with relatives on Father's Day; they were on Route 58, near Ashland, about a minute away from their old home, when they saw a cat walking slowly in the opposite direction. Russ jumped out of his truck and yelled Jasmynne's name and,

"She just came running right over to him," Debbie Brown said. "She lay on the grass and rolled and was as happy as she could be."

Joan Kortes of Harrison, Michigan, writes about when she was a young girl and the next-door neighbors moved to a new home about 170 miles away from where they lived in Detroit. Some three months later she and her mother saw the neighbor's cat Smokey back in the yard next door. They took him in—"He was skinny, matted, and there was blood on his feet"—and called the former neighbors, who immediately drove down to collect their cat that had disappeared a few days after they moved into their new house.

Another long-distance cat journey was reported in a newspaper in September 1996 about a cat named Camila who had gotten lost while on a camping trip with her family in northern Portugal. Some weeks later she found her way home, traveling a distance of 125 miles.

Susie Gay of Smithfield, Virginia, who still lives with Midnight, now nineteen years old, wrote to me about her cat's incredible journey when they lived in Germany and moved to a new home in 1989. Midnight took off, even though he had been confined in the new home for three weeks, and traveled some fifteen miles, crossing several dangerous highways and somehow crossing a wide river to reach his old home nearly two months later. He was in fine shape, being an adept rabbit-hunter.

Midnight, whose incredible journey in Germany made the press, relaxes at home.
Photo Susie Gay

Paquita Rawleigh, now living in Granbury, Texas, grew up in Spain and had a kitten named Perico. She writes to me that she was devastated when her father sold the cat to a visitor. But two months later, Perico was back home after a twenty-four-mile journey. This time, the amazed and chagrined father allowed her to keep Perico.

I have several other letters about "homing" cats. Sixteen-year-old Mittens from Michigan took two days to cover the three miles from his new home to his old one. Twelve-year-old Riley, from the same state, was enroute to be euthanized at the veterinary hospital because of a gangrenous leg. He escaped from the car and found his way fifteen miles home. Amazed at his will to live, his owner put him into intensive care, and Riley enjoyed another seven years of life. Top prize for tenacity must go to V.C. (which stands for Vacuum Cleaner, because of the way he ate), who went eighteen times from his new home to his old haunts two miles away, eventually dying from feline viral leukemia, most likely contracted from an infected cat on one of his home runs.

Not to be outdone by these super-cats, a nine-year-old Dalmatian dog by the name of Mountain Pretty Face made the local news in Grass Valley, California, where he jumped out of his owner's truck during a shopping stop. The distraught owner, Dave Almassy, camped out and searched for days before giving up on his beloved dog. Forty days later, the dog surprised him as he was watering a plant beside his trailer home near Rollins Lake; the joyful but now-skinny and exhausted dog had crossed some fourteen miles of rough desert terrain to get back to where he belonged.

Science has not yet demystified psychic trailing, an animal's uncanny ability to locate its owners in a place to which it had never been before, sometimes traveling hundreds of miles. This phenomenon has been well documented especially in the United States and objectively investigated by the late Professor J. B. Rhine of Duke University, a world-renowned parapsychologist.

A psychiatrist friend of mine told me of a German shepherd he was given as a boy by neighbors who had moved to live on the other side of New York City. The dog ran away after they left; within a few days, they found him in their new neighborhood, a part of the city he had never been to before.

There is one documented case of a cow that was sold at an auction in England and her little calf, which went to a different farm. The cow escaped from her new farm home and was found the next morning many miles away—on the farm where her calf had been taken and where she had never been before.

# The Empathosphere

I have a rational explanation for this phenomenon: it is neither metaphysical nor mystical, but has been explicit in many spiritual teachings. Simply put, we are all connected psychophysically with the sun, moon, Earth, the stars, and with each other through the realm of the senses and the emotions. It is the emotional connection with his owner or family that forms a point in the space-time continuum that enables the animal to reorient from his home base and find his family. I propose that the animal's internal sun-time clock and geomagnetic compass are used, like a directional feeling-sensitive antenna, once the animal has aligned himself toward the emotional field of his owner/family. This field, which I call the *empathosphere*, makes the space-time continuum a unified field.

Albert Einstein theorized the existence of this unified field but failed to express it mathematically. It is doubtful if the subjective element of being and emotion can ever be expressed in objective mathematical terms. The existence of this field—in which all things are interconnected and interdependent—has been demonstrated by the modern sciences of ecology and quantum mechanics.

Animals have demonstrated the existence of this unified field of being in another way. I have received many letters from readers of my syndicated newspaper and magazine columns about their pets reacting "psychically" to the death of a companion pet or human family member. A typical example is of an old Siamese cat who suddenly began to cry loudly in obvious distress at around 10:00 a.m. The veterinarian called the cat's owner about an hour later to inform the woman that her beloved German shepherd, which had been admitted to the veterinary hospital early that morning, had expired on the operating table at 10:00 a.m. The coincidence of this event with the cat's distress at exactly the same time is the kind of phenomenon that we generally refer to as *psychic*.

I have heard from other cat and dog owners whose animals have suddenly become agitated, distressed, or scared at around the same time that a human companion or close friend had died.

I have also received accounts from people reporting that their cat or dog becomes excited at around the time their spouses are due to arrive home. The animals have no specific cues, such as the sound of a footstep, the distinctive noise of the car engine, or a set time of arrival. These animals seem to be able to *feel-see* their human companions when they are some distance away, or just leaving their place of employment.

For example, a veterinarian's dog sensed when his owner closed the hospital for the night and took a fifteen-minute walk home. The animal doctor's wife never knew the exact time of her husband's arrival, but the dog would always wag his tail and go to the door in anticipation of his return. When the vet's wife started taking note of the time, it coincided precisely with his closing of the hospital door, fifteen minutes before he was at the front door of their home.

## An Ability to "Feel" Across Time and Space

I have a partial explanation for this "psychic" awareness. It is based on the notion that animals, when emotionally connected to each other and to their human companions, can "feel" across time and space and sometimes sense another's activities and emotional state.

This empathetic connection is the feeling-world, the *empathosphere*, of the animal kingdom from which most humans have become separated. Indeed, some people have become so alienated from this realm that they actually doubt that animals have any feelings at all.

As for those cats and dogs who are able to engage in what is termed psi-trailing, again I believe the empathosphere theory, which I will discuss shortly, seems to explain the phenomenon.

Those of us living in Western industrialized society are unfamiliar and out of touch with these supra-sensory powers: the animal powers that our shamanic ancestors understood and utilized. These powers also include cats' and dogs' exquisite sense of smell and superior auditory, visual, and earth-magnetic (or geokinetic) sensibilities.

Our senses are beginning to atrophy, to close down. Aboriginal people of high order and repute in precolonial Australia used these powers to heal and to live in harmony for health's sake—both personal health and the health of the environment, which they considered inseparable.

Today, these so-called primitive people have "psychic" or supra-sensory abilities. They are able to attune themselves to the unified field of being—a state they refer to as the *dream time*, and, for example, feel-see the presence of a particular animal over the next hill, or a kinsman coming toward them from a particular direction who is still a two-day trek away.

These examples of animals' higher powers should make us not only question the nature of reality as we perceive it physically and emotionally; it should also make us realize how sensitive and aware animals are and lead us to question the morality of certain practices. For example: killing animals for their fur and meat; raising them in factory farms and zoo prisons; keeping them in small cages in laboratories where they are made to suffer in order to find cures for diseases (most of which we bring upon ourselves).

If animals are emotionally connected, in some kind of sympathetic resonance with us—as witnessed by the phenomenon of psychic trailing—then we should consider how we might be affecting the entire animal kingdom by our state of mind that collectively permits so much animal exploitation and suffering, wholesale destruction of nature, and the pollution and poisoning of the environment. Since everything is connected in this unified quantum field, we might also consider the adverse consequences to us—medically, psychologically, and spiritually—of what we do to animals on the illusory grounds of human necessity.

The medical and psychological benefits of a positive, loving attitude toward animals have been proven: they help heal us. Scientists have shown that petting an animal lowers one's blood pressure and reduces the incidence of coronary heart attacks. The medical, psychological, and spiritual benefits of simply being in the country, reconnected with nature, are also being more widely recognized.

As we lose touch with nature, the animal kingdom, and even with each other, we lose these abilities and their associated sensitivities. Our survival may well depend on all of us reawakening these abilities, rather than upon new medical miracles and technological fixes. We are part of a far greater miracle that beckons us to have reverence for all life, which Albert Schweitzer saw as our ultimate healing revelation and ultimate solution to the ills of society, body, and mind. As the book of Job advises, "Listen to the animals . . . for they shall teach thee." A good beginning would be to empathize with the present holocaust of nature and the animal kingdom, to stand up and defend the rights of all animals—domestic and wild, captive and free—to achieve equal and fair consideration. Their being—their behavior and "psychic" abilities that affirm our biological and spiritual kinship with them—surely make this the first ethical imperative of a more enlightened humanity.

# 8
## | Entering the Deep Heart's Core: The Empathosphere |

Regardless of my training as a scientist—with a doctoral degree in science as well as another doctoral degree in the faculty of medicine from the University of London, England, plus a veterinary degree from the Royal Veterinary College, London—I have been able to keep an open mind to things that cannot be weighed and measured, objectified and quantified. Such subjective elements are feelings, beliefs, intuition, and the inner mysteries of life and consciousness. A scholarly approach—and by that I mean one that is impartial and unbiased, rather than one with a scientific or religious bias—is needed for when dealing with issues that are in the realms of the spiritual and the metaphysical. When it comes to evaluating animal prescience and remote sensing—what is commonly regarded as psychic communication or clairvoyance—an open mind is called for. Let the facts speak for themselves, from which we can draw our own conclusions and upon which we can then variously construct or deconstruct scientific and philosophical hypotheses and theories, religious doctrines and dogma, and test our personal beliefs.

In my earlier book *The Boundless Circle*,[1] I first raised the probability, on the basis of my own observations and the anecdotal data of others, that all living beings have an innate sympathetic connection to each other through

---

1. Michael W. Fox, *The Boundless Circle* (Wheaton, Ill.: Quest Books, 1996).

our emotional consciousness—our awareness of how we feel toward another living being, human and nonhuman, animal and plant. This connected state of awareness, in harmonic resonance with other living sentient beings through empathy, is a state of being that forms what I call the *empathosphere*. This is a sphere of being that we all share and essentially participate in, like the physical atmosphere. I proposed that humans and all living beings are in a state of sympathetic resonance. This hypothesis is supported by the sensitivity displayed by many animals of different species toward us: how they avoid those of us who are afraid or would harm them; how they approach or do not flee from or attack those who have a deeper equanimity radiating toward them.

In other words, our emotional state, and how we perceive, react to, and value animals—and each other—is communicated in this empathosphere, with profound consequences in terms of our own mental and spiritual well-being and the well-being of others, leading to euphoria or dysphoria, mutual harmony and well-being, or conflict, stress, and distress.

Cats have long been thought to possess empathic, psychic powers, although some of their abilities, such as their homing ability and early response to coming earthquakes and tsunamis, can be partially explained on the basis of physical sensations and physiological and behavioral responses. But there can be no immediate physical sensations associated with the following reactions in dogs and cats; and with our present limited scientific knowledge we cannot ascribe any known physiological process to explain how animals can react to an event occurring at the same time as they react to it, *but in a totally different place.*

Take, for example, Edna L. Thorstensen's letter from Hollywood, Florida, about her father's cat, whom she was allowed to take to the hospice unit of the hospital to visit her father. "The last night we were there I knew my dad would not be with us too much longer. That night when kitty and I came home, she started running through the house howling. I had no idea what was wrong with her. A few minutes later the hospital called and said that my dad had just passed away. She knew it. It took her a long time to get over my father's death."

A letter from Angela in Minnesota tells of her husband, who was dying in a nursing home and who they visited daily. "One night around midnight our eighteen-year-old cat, on my bed, gave a strange sound that woke me up and I sat up. A second later the telephone rang telling us to come to the home. The cat knew just when the Lord took my husband."

Kathy Rector from Mellenville, New York, writes: "My husband's grandfather found a stray golden retriever and named her Penny. They were inseparable for many years, until he went into the hospital. One day Penny began to howl and howl, and my grandmother knew her husband had passed before she got the phone call from the hospital a few minutes later."

According to Stephanie N. Abdon from Winston-Salem, North Carolina, her grandmother used to pet-sit for various animals. "One of those pets was my brother's dog Dixie," she writes. "When my mother called to let my brother know that Granny had passed away, he replied that Dixie had already let them know. Apparently Dixie had come into the bedroom and began to moan, as if she were mourning. This is apparently the only time she whimpered in such a way. I think this is a perfect example of your empathosphere theory."

Karen Beloncik from Schenectady, New York, gives a similar account. She was taking care of her uncle's boxer dog, Champ, while he was in the hospital. "One evening," she writes, "Champ was sound asleep when he suddenly woke up barking and running back and forth throughout the house. A short time later, I got a call telling me my uncle had died. When I found out what time he had died, I realized it was the exact time that Champ had been carrying on. I truly believe that Champ knew what had happened."

Delyne E. Eddins from Joshua, Texas, remembers an event that occurred when they lived in Nebraska. Her family dog had become very attached to her grandfather, whom they visited in Kansas for a couple of weeks every summer. She writes: "One day my mother received word that her father was very ill, so she went down to Kansas to be with him. One day the dog started acting very strange and was crying, howling, and acting like he had lost his best friend. The remark was made, Do you suppose grandfather died? A few hours later my mother called and said her father had passed away. We asked what time and she told us. It was the same exact time that the dog was crying and howling."

Bay was a mutt who was adopted by the uncle of Cindy Weldon of St. Paul, Minnesota, when he was serving in the Peace Corps in Antigua, West Indies. Bay's family included Cindy's grandfather who, at 101 years of age, was flown to Baltimore for heart surgery, accompanied by her uncle. Her aunt stayed at home with Bay, who "kept a vigil by grandfather's chair by day, and by his bed at night. He would not eat. My aunt had to carry him outside to go to the bathroom. This went on for five days. My aunt was worried Bay might die. My grandfather died about one in the morning shortly after surgery. That morning Bay gave up his vigil. He ate and went outside on his own. Somehow he knew that Grandpa had been released from this world. Bay has been his usual self ever since."

Sometimes the animal's prescience has a happier outcome, like that of a little black mongrel dog named Bubbles, who pined for months while her beloved master George was posted abroad during World War II, "taking prolonged naps in his bed," according to George's niece Candy Killion from Davie, Florida. She writes, "When my grandmother got mail home from the front, the dog would go wild; apparently, she picked up his scent on the letters he wrote. But strangest of all: as the war was ending up, one morning my grandmother heard the dog pacing back and forth, whining and crying at her front window. There wasn't a thing in sight to provoke the animal. This went on about a half hour until a city bus that stopped at the corner made its next drop. When the bus came into view, Bubbles went completely hog-wild, barking and scratching at the window. When my grandmother went back to scold the dog, she nearly fainted. George, in his uniform, was walking down the sidewalk with his duffel bag. No one knew he was coming—except the dog."

One sad reading of human behavior by a companion animal was shared with me in a letter by Dianne Payne, Palmetto Bay, Floria. Her husband, a veterinarian, suffered from depression, especially toward the end of his life. "A couple of weeks before he died he told me that he didn't know what was wrong with Lacey [their Labrador retriever], that she wouldn't 'come' to him anymore like she used to. Although she would still wait to greet him at the door, once he came in she would retreat under the kitchen table. My husband unfortunately took his own life and I have always wondered if Lacey 'knew' that he was going to do that. I have no other explanation for why she started distancing herself from him in the two weeks prior to his death." Ms. Payne described a regular ritual she had with Lacey in the evening, saying "Daddy's home" when she heard her husband enter the garage. Lacey would always run to the family room window, look out and bark, then run to the door with a "present" in her mouth and wait for him to come in. Soon after her spouse's death, Ms. Payne once tested Lacey's response when she said the familiar "Daddy's home." Lacey dropped her ears, did not go to the window, and instead looked sad and went and lay down. The concern shown by animals toward suffering human companions reveals the depth of their empathy.

Mary F. Wilson from Livonia, Michigan, writes with reference to animals sensing that their owners have died or been injured: "I am a hospice nurse and have been told of many of these instances by patients' families. So often in fact, that I do not consider it unusual at all. Frequently families will tell me that a beloved pet will suddenly begin staying very close to the patient, even though we can see no outward signs of a change in the patient's condition. Subsequently the patient usually declines and dies within the next few days. I also am told occasionally that after the patient has died, the animal will avoid for some time going into the room where the patient had resided. I think that this demonstrates that animals are much more 'tuned in' to physical changes in 'their' people, and also shows that animals grieve."

Mac, a Scottish terrier, worked as a therapy dog with Larry Underhill from Norway, Michigan, at the local Veterans' Administration Hospital. Going the rounds of the beds on Saturday, they came to the shared room of one of Mac's favorite patients, but his bed was empty. Larry decided to take Mac in anyway to visit with the other patient in the two-bed room. "As I started into the room," he writes, "Mac would not go. It was like trying to pull a black brick someplace that he did not want to go. Absolutely, no amount of coaxing or treat-offering was going to change his mind." Soon after a nurse told Larry that the patient had died just the day before. "Chills started to run down my back," he writes, "when I realized Mac had seen-felt something that was there. Maybe he felt the spirit of his buddy was gone on, I really don't know. However it was an experience I will never forget. Unfortunately within six months, we discovered that Mac had cancer and we decided to put him down. All his buddies at the hospital were saddened when they heard the news."

Roxanne MacKinnon sent me an interesting letter about her mother, a cancer victim passing on peacefully at home, and how her mother's dog reacted. Roxanne writes, "On the day she expired, her miniature schnauzer, Sarah, was lying on my husband's lap in the living room across the house, while he read the paper. I was in the family room watching my mother take her last breath. Suddenly, Sarah jumped off my husband's lap, ran into the family room, jumped up on the side of my mother's bed, and looked at her as she had taken her last breath, then looked up toward the ceiling, as if she could see or sense the soul leaving the body. Sarah then proceeded to get down from my mom's bed, and ran back into the living room to once again be with my husband. I have *never* seen anything like this before. Sarah came out of nowhere the minute my mother took her last breath."

The vigilance and empathy of animals may even cross over the space-time continuum as in this remarkable account by Lois R. Smithwick from Hurst, Texas. On a cruise with her husband, she woke one morning at 5:00 a.m., hearing their dog, Radar, scratching at the door. (Radar is her daughter's beloved rescue dog who was going to be killed by his former owners because they were moving into a new house; she and her husband took care of the dog each day while their daughter worked as a nurse.) "It was as natural as if we had been at home with Radar getting my attention to check on 'Papa.' When I first heard the scratch my first thought was that I was missing him so much I was hearing things. Then when I became fully awake I realized all was not well with my husband. When I checked his blood sugar it was 27. If it gets below 50 it is said to be critical or the patient can go into a diabetic coma. I was able to go into immediate action and give him glucose to raise his blood sugar. When I think about my normal wakeup time I get chills down my back. It is mind-boggling to me to think that Radar's thought processes got to me when he was in Hurst, Texas, and we were aboard a cruise ship in the middle of the Mediterranean."

Radar (what an appropriate name!) and the other wonderful animals in the above testimonies affirm that animals do indeed possess super-senses of prescience, deep empathy, as well as the ability to sense *and* communicate remotely. This lays the ground for launching into the next chapter that explores the profound realm of animals' "extra-terrestrial" or after-life communication and manifestation from beyond the grave, when beloved animals no longer exist on this mortal plane to share their lives with their human companions. But still they do, and somehow can. The evidence to be presented raises profound philosophical, religious, metaphysical, and other challenging questions that some readers will find life- and personal-belief affirming. Others will find it unsettling and, I hope, life-changing—in terms of their regard for and treatment of animals sanctioned by the dominant culture's prevailing worldview in these times.

# 9
# | Animals Communicating after Death: The Evidence |

Humans, being aware of our own mortality, have elaborated various beliefs in an after-life, which are part of every culture around the world and evident in the earliest recorded history of our species. But is it all hocus-pocus—the notion of life after death, of life after life, of some heaven to come?

It is all a simple fabrication of the human psyche to cushion the fear of death and the feeling that, if there is only this one life for us, then what's the point of it all? And why bother even trying to be good, if after this life there is absolutely nothing—nada, zilch? People who conclude a life after death is a fabrication through what they call "reason" are in my mind going beyond reason to embrace nihilism or nothingness.

I believe the significance of your present life is connected to your ancestral spiritual past and future—from grave to grave, womb to womb, world to world, and universe to universe. Australian aborigines call these spiritual connections through time and space our *song lines*. Christian mystics speak of the transmigration of the soul from life to life as the journey of the pilgrim soul, a vision shared by Hindus, Buddhists, and many others. This notion of the traveling of our souls or spirits from life to life, called reincarnation, was declared anathema—wrong thinking and bordering on heresy—by the early Church of Rome (for political reasons, most theological historians believe).

In Western civilization, the Age of Reason—as it is called, around the sixteenth and seventeenth centuries—further eroded earlier spiritual beliefs by declaring that only humans and not animals have immortal souls, because animals are our inferiors, lacking in reason and morality. This chauvinistic view of animals is alive and well today in many circles, from those involved in exploiting animals for profit and those seeking to protect the status quo of human superiority and domination over the rest of God's, or Mother Earth's, creations.

Now just suppose that dead animals could communicate and even manifest themselves physically (or, according to theosophists, as an etheric double or astral body) to their human companions still living on this plane but grieving their departure terribly: What if this communication and manifestation could be verified? Would this not upset the apple cart of the nihilists, rational materialists, and those who think only humans have immortal souls and are special? The impact would be as profound as the arrival of intelligent life forms visiting us from outer space.

As for the evidence that I have gathered of the persistence of animals' spirits after their death, the following letters provide irrefutable proof. First, there is the partial manifestation and communication with the human loved one in the form most often of familiar touch. The veracity of this after-death communication is reinforced by the fact that several people have independently experienced *the same thing*. Second, further affirmation of after-death communication—and therefore existence of the animal's spirit after physical death in this dimension—comes from more than one individual seeing the manifestation of their beloved animal, partial or complete, together at the same time, or separately at different times in the same place.

No matter how much parsimony, skepticism, scientific objectivity, and impartial judgment I apply in assessing the following accounts of "visitations" by deceased companion animals, the accounts speak for themselves. They give no intimation of some mental creation and projection, or associative conditioning, in which the person claiming to have experienced such a visitation/communication is simply hallucinating or imagining that their beloved animal has returned from beyond the grave in order to help make them grieve less from the loss. While such manifestations are clearly comforting and affirming of the belief in life after life, I find no evidence in these letters that the people involved "called up" their deceased animal companions to comfort them. Rather, they appeared spontaneously and unexpectedly, thus ruling out the possibility of psychic conjuring and imaginative creation. (But this is not meant to imply that the deceased, human and nonhuman, cannot be "called" in times of need through prayer, meditation, ritual, and dream states.)

Many people believe there is a Heaven and a Hell, and that bad people go to Hell, while angels and our beloved animals (whom some people regard as angelic beings) abide in Heaven. But I like what Milton wrote: "The mind is its own place, and in itself can make a Heaven of Hell, a Hell of Heaven." I believe that the notion and creation of Heaven and Hell are purely human in origin.

The love of animals reveals to us the boundless love of the Creator, just as our love of nature reveals to us the nature of love. The thirteenth-century German theologian Meister Eckhart wrote: "Apprehend God in all things, or God is in all things. Every single creature is full of God, and is a book about God. Every creature is a word of God."

Saint Francis of Assisi said that it is through animals and nature that God is revealed to us. Hence he was recently named the Patron Saint of Ecology by the Catholic Church, but has been long and better remembered as the Patron Saint of the Animals. His teachings clearly influenced Eckhart. If every creature is a word of God, then God is speaking to us through them, which is precisely what St. Francis experienced in prayer, meditation, and rapture as he preached to the birds of Assisi in praise of the Creator and all of creation.

In light of this, it shouldn't be surprising if animals could communicate with us after they have entered the Light and left our physical presence. And indeed they do.

## Experiences of Those Who Have Lost Animals

Elinor Lovegrove, from Shelton, Connecticut, wrote to me about an experience she shared with her husband. She wrote in a letter: "My husband had a grey cat named Rosemary. We had her for seventeen years and she was a dearly beloved pet. When she died my husband was broken-hearted. We were sculptors together and he spent a lot of time in our studio where Rosemary sunned herself daily on a drawing table. A week after her death he called me to the next room. He asked if I had changed my cologne. As I went to the doorway, I was struck by a lovely fragrance. I said, 'What on earth is that?' He said 'I don't know.' It was suddenly gone. It was February, the windows were shut. We both puzzled over it but had no answers. That night we were watching TV in the living room. John shouted 'Ellie look!' The fragrance filled the room and his darling Rosemary was leaving the room with her tail up in the air! This was normally her sign of great satisfaction. Suddenly she was gone and so was the lovely fragrance. We both saw her, in living color. John was in tears. This episode comforted us."

That fact that two people saw the same manifestation of their deceased animal companion in the same place and at the same time certainly supports that this experience was not simply a projection from John's imagination. They also smelled the perfumed aura of Rosemary's spirit, or essence-nature, and saw her physical manifestation that could well have been willed by Rosemary as a final act of love to assuage their grief and give them every assurance that she was indeed well and in the best of spirits in her new dimension.

Another complete manifestation, this time of a dog, was reported to me by Marion V. Russo from Hendersonville, North Carolina. She writes: "It was ten days after Noelle, our German shepherd, passed away. It was dusk and my husband and I were on the deck when a large brown dog with black markings went running past us. We gasped in shock and we said to each other, 'Did you see that?' It was Noelle. We could not speak for quite a while. I said she came back to say a final farewell. My husband did not want to admit to what we both saw, but it was our dear Noelle."

Ms. Russo's second dog, another rescued stray that was part shepherd, also manifested herself after death. "After Chrissy died I saw her vision and heard her footsteps in the house for many months. I still hear her, especially when I am in the kitchen. She loved her food and when my husband has a late snack he hears her footsteps too. In fact he looks down at the side of his chair and says, 'Chrissy, I know you are there.'"

Some photographers have claimed to have captured images on film of alleged apparitions, which may be fake or mere artifacts of light or defective processing. But one photograph sent to me by retired U.S. Air Force officer and experienced photographer Robert J. Young of Kernesville, North Carolina—which he had examined by film experts—may indeed have captured a genuine manifestation. He was in his garden with the sun behind him one early-May afternoon and took two pictures of his dog Cheyenne using a Minolta single lens reflex camera and Kodak 800 film. The first picture showed Cheyenne lying beneath a Colorado blue spruce tree that had been planted in memory of their deceased dog Lady, who had passed on in November the previous year. The second picture revealed a light, transparent cloud hovering in the air in front of the tree just above Cheyenne.

Mr. Young wrote to me to say, "God would not call home one of his creatures and leave behind those that loved the dog without some way for them to know that the dog was safe and in his loving care. And that the dog although gone was still watching over the people she loved. I believe that He allowed our beloved Lady to visit us and let us know."

Judith A. Sellins, who lives in Westerly, Rhode Island, often sees her deceased black cat Whiskers. "But I haven't been the only one to see him," she writes. "My daughter as well as others have also seen him, asking me when I got another black cat. But I don't have a black cat. He has been seen walking into my bedroom and through my dining room."

Cathy Andronik from Bridgeport, Connecticut, sent an interesting account: "For thirteen years my family had a pet cat, Rusty, who with his temper was quite a challenge to love, but who could be affectionate when he chose. After a long, good life, he was euthanized well over ten years ago. One thing he loved was to jump on my bed early in the morning and stare into my face, nose to nose, until he was satisfied that I was awake. One night last fall, I was awakened by the sensation of something about the size and weight of a cat—landing on my bed. When I opened my eyes, I clearly saw the shape of a cat's face and ears in front of me, and sensed an animal's presence. The shape was so solid it partially blocked the numbers on the alarm clock. When I blinked a moment later it was gone. While I remember Rusty frequently and fondly, I am certainly no longer mourning his loss. He was, however, the special companion of my mother, who passed away about three years ago, and I like to think that he paid me a visit to let me know that he and my mother are keeping one another company again."

Donna Ballard of Flint, Michigan, was on a trip to Greece and left her husband in charge of their cat Beau, a Maine coon, who was in failing health. She writes: "The Thursday before I was to come home I was reading in bed at three in the morning when I noticed a movement in the corner of the room by the ceiling. I looked up and there was Beau. He floated down to my book and disappeared. I did not go back to sleep so I know I didn't dream it. On my return, my husband picked me up at the airport and I said, 'You had to put Beau down didn't you?' He answered, 'Yes, how do you know?' 'Was it Thursday?' I asked. 'Yes, it was Thursday,' he replied."

Cheryl Morgan from Milford, Connecticut, wrote to say that she is a devout Catholic and all three of her beloved dogs had to be euthanized at different times because of intractable old-age health problems. First, after her Border collie was put down, she says, "As I was riding my bicycle, I could see him running alongside me in spirit. He was very happy, and he was like a puppy again. This enabled me to let him go."

Subsequently, after her beloved beagle was euthanized, she saw the beagle and Border collie running and playing alongside her bicycle, and she saw all three of her dogs running beside her on the bicycle and happily playing soon after she had the third old dog, a cocker spaniel, put to sleep.

Anita Perry from Anaheim, California, was regularly "visited" by her dog Barney a month after he passed on. When still living on this plane he used to bump the side of her bed with his nose when he wanted to go outside, and she was awakened by the same bump, which she initially thought, living in California, might be an earthquake. These nightly bumps went on for many months. "One night when I awakened to turn over in bed I saw his reflection in a mirrored closet door. I was surprised but so glad that he was

Beau in life, whose spirit traveled around the world to contact his beloved human companion on the day he died.
Photo Donna Ballard

there. Looking closer at the floor, in front of the mirror, where he should be, he wasn't there. He could only be seen in the mirror."

Some time later she asked Barney's veterinarian if he had heard of anything like this before. "He laughed and said that his dog Charlie, who had died ten years before, is felt jumping off the bed every night by both he and his wife."

Ms. Perry has collected similar accounts of the "visitations" of departed animals, noting, "The owners of those pets said their life was changed for the better by their experiences," and that such documentation would greatly help in grief support for those who have recently lost a pet. I would add that this will also help those who are *about* to lose a loved one, human or nonhuman, or who are about to die themselves. Sharing some of these accounts with my old and terminally ill father-in-law Jim Krantz, a salty World War II veteran tank commander who has seen enough of suffering and death, elicited a sage response: "Animals know much more than most people give them credit for."

Alice Cruze in Holyoke, Massachusetts, was visited by her deceased ten-year-old Peek-a-Poo Austin, about a month after he was put to sleep: "I was slightly awakened by something jumping onto the bed. I recognized the familiar walk and knew this wasn't possibly Austin. Yet I reached out my hand and felt him begin to quiver all over, something he did when excited in his healthier days. I felt him kiss my cheek. I ran my hand down his back. I needed to feel the tail as he had a distinctive curly tail. Just as my hand touched his tail, he was gone. I felt my cheek and it was damp. The next morning I told my husband about my 'dream' and he said, 'I thought I heard Austin's footsteps in the room last night but realized it couldn't be.' I truly believe Austin came back to me to let me know he was happy and in a good place."

Sometimes the manifestation of the animal is only partial, like Marian West's deceased stray cat, Whitey, who lived with her in St. Louis, Missouri. One evening while reading in bed she felt, "a cat jump onto the bed and begin to knead my leg. Without looking, I reached down to pet him and was surprised to find no cat on the bed. The stray loved to sleep curled around my legs and would always go through the ritual of kneading before he settled down to sleep. Since then he's visited me several times—I would see a flash of white from the corner of my eye. My other cats have acted strangely from time to time and in my heart I truly feel Whitey is popping in to say 'hello.' I might add that these visits rather than being frightening are quite comforting."

People who have experienced incomplete manifestations of their deceased companion animals have variously heard them breathing, drinking, eating, and walking on the floor, and they felt them close, especially lying beside them on the bed or jumping onto the bed. Sometimes they appear fully in a dream that can be so vivid that the dreamers may believe they were actually awake.

Awake or dreaming, Aaron Marten was in the Navy on board ship asleep in his bunk when, according to his mother Louise Marten of Galveston, Texas, he was awakened by a scratching at his door. "When he opened it," she writes, "there was Lucky, his dog. He jumped onto Aaron's bunk and Aaron went back to sleep thinking he was dreaming. In the morning Lucky was gone; but there were bits of white fur on his blanket." When Aaron was on shore leave he discovered that his beloved dog had died around the time he had the visitation on board ship.

Some people feel their attachment and grief after their animal companion has died may keep the animal from passing on—and they should be allowed to be "released."

Lisa Biagiarelli from Easton, Connecticut, had a difficult time getting over the loss of her old cat Bobo, whom she often felt was around her feet and was staying around to comfort her. So she spoke to Bobo and said: "'I know you are here with me. I know all you ever wanted was to be with me. But if you have the chance to go someplace better, to be with St. Francis (to whom I used to pray for Bobo's health), you should go. I will be OK—you can go.' I immediately felt him leave me, and I have never felt him under my feet since that time. I know he is in a better place and I also know that someday I will see my beautiful Siamese again."

Alice Weir was sitting in her armchair beside the window in her home in Shelton, Connecticut, grieving over the loss of her old cat Chelsan. He had a habit of jumping off his bed on the bay window onto the left arm of her chair to get to the floor. "As I sat crying and thinking of him I felt a brush like fur on my cheek on the left side of my face where he jumped. I felt it was his way of telling me thanks for the good care I gave him before his passing and letting me know he was alright now."

Maryann Gallant in Stuyvesant, New York, shares a similar experience. Her mini-poodle Pepe started going blind at age eleven; until he passed on six years later, he developed the habit of following her around the house, staying close to her right leg and repeatedly touching her right calf with his cold nose. She and her husband grieved for Pepe for days and days. "One

Saturday morning," she writes, "I got up and without a single thought of him started to walk down the dark hallway to the kitchen. All of a sudden, I felt a cold wet nose on my calf. My heart jumped for joy and I looked down expecting to see him, but alas there was no Pepe."

I have received accounts of surviving animals seeming to respond to the spirit-presence of a recently deceased animal companion. Such an experience was shared by Ann Simanton from Houston, Texas, with her young cat. The evening after she had to euthanize her old cat she went to bed. The old cat had always slept with her on her bed. This evening she felt his presence in his usual place, on a bundle of covers that, "started moving up and down and vibrating as though it was breathing and purring." She checked to see if the younger cat was under the bundle, but she was asleep on the couch. Going back to bed, she put her hand on the bundle. "It was moving and purring, and my hand and the bundle was covered with small 'sparks'" she writes. Then, "I heard a scream from the other room and the younger cat, who had not yet set foot in the bedroom since the older one's departure, came

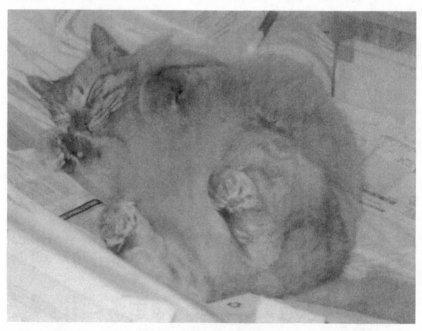

A roly-poly Spencer in life, whose after-death manifestation was seen at the same time by its owner and a surviving cat.
Photo Ann Simanton

screaming into the room, jumped on the bed, ran to the sparking, purring bundle of covers and started kneading it like mad and 'talking' to it, in the little 'chirpy' language they used between themselves." This was the only visitation she had from her deceased feline companion. Interestingly, the next day when Ms. Simanton described this experience to her veterinarian, the vet confided that she had seen each of her two dogs after they had passed on, but had never shared this experience with anyone before.

Karin D. Welsch from Albany, New York, was a student whose parents delayed telling her of the death of her cat Socks until her final exams were over. The night they told her she had a dream: "Socks came to me; he looked at me and I was made to understand that he had returned to give me the one last visit I had been wishing for. The images in my dream were so vivid and the message so clear that it has forever affected my views on life, death, and afterlife."

\*   \*   \*

Many people are uncomfortable sharing such experiences—fearing the ridicule of skeptics and the prejudices of rationalists and materialists, as well as those who do not believe in disembodied spirits and in animals being living souls, just like you and me. Others cast doubt on their own spiritual and metaphysical experiences, dismissing them as products of the imagination to help relieve grief, guilt, anger, loneliness, and other feelings associated with the loss and mourning of a beloved animal companion.

But as long as we deny the deep heart connection these animals can provide, we will continue to deprive ourselves of experiencing one of the greatest, life-changing affirmations of universal love that is manifested through the devotion shared between us and our animal companions. They are not simply our property; they are part of life universal in spirit, and part of the great mystery of the soul's journey from life to life.

It is easy to dismiss the miracles of every day and the intimacy with all that is sacred, which is invited by an open heart and mind, as the animals teach us time and time again—life after life. Many people have been profoundly moved and comforted by these after-life communications from their animal companions, and their lives have been significantly changed by the revelation that there is more to mortal life than we know. We must all agree with Jim Krantz that animals do indeed know far more than most people give them credit for.

# 10
## | Delving Deeper: Cats as Our Mirror |

### First Impressions

Animals often respond to us in ways that make us think that they must be psychic, or are reading our minds. But they are picking up unconscious cues from how we perceive and react to them, and are therefore mirrors, to varying degrees, into our own subconscious. This mirroring is also a reason, I believe, why animals often come to closely resemble in temperament, if not physically, the people with whom they live. When we are sad, depressed, insecure, emotionally disturbed, or relaxed, happy, and playful, so too, most often, are our animal companions. They are not only empathetic, feeling and reacting to how, if not to why, we are feeling; they are also captives of our emotional and physical environments that we share with them. People who are indifferent toward animals generally think animals are indifferent toward them, just as those who are violent are recipients of animals' fear and sometimes anger and aggression. A greater appreciation, therefore, of how we affect animals for better or for worse can do much to enhance our own understanding of how we are seen by others, human and nonhuman, and how our intentions and actions are understood by others.

In my experience, our first impressions of animals and the reactive feelings we recall from our earliest childhood memories determine our ability to understand and communicate with animals as adults. No matter how good our empirical sciences may be in this regard, our feelings and attitudes—our deepest beliefs, fears, and longings—ultimately determine how well we understand and communicate with each other, and therefore by extension, with other animals and nature.

My earliest memories of seeing other animals and of being with cats are of awe, wonder, and delight. They were something other—profoundly other. It was their diverse "otherness" that helped me realize the world was not just for people but for myriad other life forms. In their diversely adaptive modes of being and consciousness, they evidenced purpose, will, and intentionality. Neighborhood cats were among my first teachers and closest nonhuman friends. Many taught me the nature of respect and to respect their natures, and that while some liked to be petted, others were more comfortable if I kept my distance, which hisses and paw-swipes and the occasional scratch soon taught me.

As a child, I found that every cat was a potential buddy, usually friendly, and always fascinating, nontalking, yet somehow an easily understood significant other. They taught me about being a cat, in terms of their likes and dislikes, and of their ways of expressing various feelings, needs, intentions, and expectations. Many adults have to let go in many ways before cats can be their teachers and healers. It's much easier for children who are not afraid and who have not yet learned the anthropocentric catechism of adults who would delude them into regarding and treating cats and all nonhuman animals as inferior, irrational, even immoral, unfeeling soulless beings.

Professor Marc Bekoff has done much to encourage other scientists to examine this catechism in relation to their own studies of animal emotions. His collection of evidence in *The Smile of a Dolphin: Remarkable Accounts of Animal Emotions*[1] posits that nonhuman animals are not only our equals, but often our superiors in the domain of feelings and sensibility. His work is extremely convincing.

But as writer Isaac Bashevis Singer noted, "In their behavior toward creatures, all men were Nazis . . . The smugness with which man could do with species as he pleased exemplifies the most extreme racist theories, the principle that might is right."

Our expectations of how cats ought to behave are often based on a lack of behavioral understanding and biased by anthropomorphic comparisons,

---

1. Marc Bekoff, *The Smile of a Dolphin: Remarkable Accounts of Animal Emotions* (New York: Discovery Books, 2000). See also Marc's recent book *The Emotional Lives of Animals* (Novato, Calif.: New World Library, 2007), and Jonathan Balcome, *Pleasurable Kingdom: Animals and the Nature of Feeling Good* (New York: Macmillan, 2006).

values, and judgments. Cats that hiss at visitors or hide under the bed notwithstanding, the ways of animals can be inspiring, educational, and sometimes highly amusing—but only when they are free to be themselves, ideally living not just with humans but with at least one other member of their own species. When living with another member of their species, their ways are not grossly distorted by an exclusively human social life that may be far from healthy—even for the human species.

As Professor Nikko Tinbergen, another Nobel Laureate and founding father of ethology, said: "Every time you have two animals together, you have an experiment."

But my best teachers have been cats and other animals themselves. When I first began to study their behavior, I made the novice's mistake of being an intense observer instead of being a seemingly indifferent bystander. My behavior made some cats apprehensive because being observed meant being looked at—and that is very threatening to some cats.

When I taught animal behavior/ethology in the 1970s at Washington University, St. Louis, I made it clear in my introductory lecture that the behavior of animals provide us a window into their minds, their consciousness, because much of their behavior expresses their motivations, needs, intentions, expectations, and especially their feelings or emotional states. In social contexts, their behavior enables us to better understand social relationships and modes of communication (or languages)—with the realization that animals understand each other and are adept ethologists themselves! Also, in various environmental contexts their behavior might reveal adaptively evolved mechanisms, patterns, and strategies. Some of these, like their communication signals, are innate or instinctual. They may not be consciously mediated or modulated (just as when we unselfconsciously and spontaneously smile at someone). But this does not mean that animals are unconscious, instinctual automatons. I call such a parsimonious "scientific" interpretation of animals' behavior *mechanomorphization*.

## The Notion of Animals as Machines

The belief that cats and other animals are unconscious, unfeeling machines has been around for several centuries. Such parsimony is part of the so-called objectivity of science and of perceiving and, unfortunately so, often treating animals as unfeeling objects of scientific investigation. This perception is based on the Cartesian philosophy—the mechanistic and dualistic philosophy of seventeenth-century French philosopher René Descartes.[2] This highly influential "Enlightenment" era thinker contended that the screams of dogs being cut open and dissected in the name of science and medical progress were simply the sounds of their clockwork mechanisms breaking down.

Descartes held that only humans had rational souls: "I think, therefore I am." He followed the earlier Greek Stoic philosophers' view (that St. Thomas Aquinas put into Christianity) that animals are not rational because they have no language. Being irrational, they lack conscience and therefore cannot be moral agents and therefore be entitled to or make claim to having certain rights and interests. Being brute and irrational, animals can have no consciousness of themselves either, according to Cartesianism. This means that when they are physically injured, like a clock, they become dysfunctional but are not conscious of any pain. Having neither reason nor language, they can have no real emotions, like fear, pleasure, anxiety, and affection. Such emotions are exclusively human. To attribute such human-like emotions to animals is to commit the Cartesian taboo of *anthropomorphization*—of regarding animals as being more similar to us than they are different. On the contrary, when animals experience sensations (like thirst, hunger, or pain) and associated conditional signals (like a whistle or a command), their reactions are to be regarded as being purely unconscious and instinctual and therefore unconscious.

People were once burned at the stake, hanged, drowned, or excommunicated for having the "witch's power" of being able to talk to the animals, and of having a way with them that other people did not possess. Such people—who were often lonely and reclusive, preferring to share their lives with many cats and other animals rather than associate with ignorantly superstitious neighbors—were seen as heretics, pagans, devil worshippers, and heathens promoting animism and pantheism. Cats especially were associated with evil; and black dogs and other animals that were black or had unusual markings—or were wild but had been rescued, healed from injury, and become tame "familiars" in the hands of the "witch"—were seen to be possessed and under satanic control.

One of the reasons why people enjoy the company of cats and other animals, and have one or more in their lives, is perhaps because the animal ✓

2. For more on Descartes and the shaping of current attitudes toward animals, see Keith Thomas, *Man and the Natural World: A History of the Modern Sensibility* (New York: Pantheon, 1983).

is an antidote to the kind of soulless, heartless world of materialism and consumerism in which most of us are ensnared—a context imbued with superficiality and a lack of intimacy. *The human spirit needs intimacy* in most human relations.

As psychotherapist and former monk Thomas Moore in his book *Original Self* concludes, "Politicians and the politics of normalcy contribute to the fostering of a soulless society, hell bent on materialism and superficiality. If you are normalized to a psychotic society, it is not a good place to be . . . You've got to have one foot in and one foot out of this society, because this is a soulless society." A life without passion and meaning is no life at all. A soulless society is a society that regards animals as soulless and treats them accordingly. Moore advises, "Do things that are counter to society. It keeps you sane. Not normal but sane. You can't keep it all inward, or you go nuts."

I believe I am not alone in acknowledging how my animals, as well as my work as a veterinarian and animal rights advocate, keep me from going nuts. My animals keep me in touch with the real, because they are authentic, empathic antidotes to a soulless society—just as those people who understand their need for cats, to be with horses or other companion animals, or to work in wildlife rescue rehabilitation will attest to.

## Love for Animals

My studies of animals have taught me how unreasonable, irrational, unconscious, and terrifyingly devoid of emotion some humans, rather than animals, can be. How humans behave toward other animals reveals much about their attitudes, beliefs, and values, notably the materialistic, the mechanistic, and the scientific detached and "objective," as well as the controlling, the dominating, the fearful, the indifferent, and, yes, the caring and the loving.

I once witnessed Professor Konrad Lorenz, Nobel Laureate and one of the founding fathers of the science of ethology, address an international assembly of academics. He said, "Before you can really study and understand an animal, you must first love it." Many Cartesian scientists at the conference did not agree with Konrad; some sitting close to me said that Konrad had "gone soft" and that such a remark was unscholarly because it was subjective, emotional, and therefore unscientific.

What Prof. Lorenz was advocating was an element of human affect that he felt was as much lacking in our everyday relationships with each other and with other animals as it was in the research laboratory and field. He told me that he was concerned that progress in the science of understanding animals through ethology, and public appreciation and concern for animals, were being undermined by latter-day Cartesians, mechanistic reductionists, and "objective" behavioral psychologists. He trusted that "I couldn't disagree less" with him—which was quite true.

It was also true that mere sentimental love was not what he was calling for. Sentimental and possessive "love," with all its projections and conditionality, is no basis for acquiring a deep understanding of animal behavior and consciousness. He was calling for a deeper, empathic connection that has its origins in our own aboriginal past where our ancestors could indeed "talk" with the animals and "become" them. While this kind of relationship with animals may seem quasi-spiritual or mystical to our modern sensibilities—and certainly taboo to latter-day Cartesians mechanists—it was immensely practical and a survival skill for the adept hunter, and later for the pastoralist and farmer.

The kind of love that Prof. Lorenz was advocating was both empathic and caring, which are the ingredients for a deeper understanding of animals' emotions and intentions, so critical to our relationship with animals—and with each other, especially our children. Such love makes for a mutually enhancing relationship between human and nonhuman. But where it is believed that animals are "dumb brutes," lacking in reason, intelligence, and feeling, there can be no such relationship. Nor can there be when it is taboo to infer certain emotional states in animals showing virtually the same behavior as we would under similar circumstances: the taboo of anthropomorphizing.

## How People Affect Animals

Cats and other animals have taught me that how they are perceived by people—as friendly or dangerous, even manipulative and untrustworthy—influences how animals respond to those people. Furthermore, people's perceptions, which are influenced by their attitudes and beliefs, greatly determine how the animals' behavior is interpreted. It is, of course, naïve to believe that all animals are harmless and friendly—or harmful and afraid or aggressive—and to then respond to them accordingly. It is no less absurd to respond to them as if they were simply unfeeling automatons or mere commodities—inferior beings that are simply objects of possession. No mutually enhancing human-animal relationship could ever develop from such a mind-set, or from the mind-set of those who regard wild animals simply as trophies or pests, or of believing domestic animals to be degenerate versions of their wild ancestors.

A mutually enhancing human-animal relationship, based on the kind of love that Prof. Lorenz advocates, is enlightened self-interest, for the dairy or pig farmer and for all who have cats or other companion animals in their homes. As the old saying goes, Where there is love there is understanding; but where there is ignorance, there is prejudice and fear. Several studies have recently documented how farmed animals are healthier and more productive when they are on friendly terms with their caretakers. Sows produce more piglets, hens more eggs, and cows more milk. Gentle, routine handling of infant animals, and also of their mothers during pregnancy, amazingly enhances their resistance to stress and to various diseases later in life.

I am surprised and saddened that many scientists are such doubting Thomases that they have interpreted the play behavior of cats and other animals in purely instrumental terms—reflecting a mechanistic orientation—contending, for example, that through play, puppies and kittens are simply learning how to improve their fighting and hunting abilities and establish dominance. There is no place in this orientation or paradigm for considering the possibility that animals play together because they enjoy it. Of course, there are other benefits from social play, not only physical exercise, but also establishing and maintaining affectionate ties and reinforcing social rank. There is also an element of creativity as animals make up new games (sometimes using inanimate objects as toys) and orchestrate variations in the frequency and duration of various actions and sequences. There are also role reversals, as when a socially dominant dog or cat allows a subordinate to playfully "kill" him or mount him.

## Cats as Mirrors

That people often resemble the animals they live with, sometimes with striking psychological and somatic similarity, especially in terms of temperament and demeanor, may be more than coincidence. A process, which I have termed *sympathetic resonance*, may account for this phenomenon, which cannot be dismissed as pure coincidence.

Cats are as much a mirror of our humanity as they are subjects of our inhumanity. Their well-being as our domesticated companions reflects our own well-being. This is determined less by our economic well-being than by our spiritual capacity to care for our families and communities, both human and nonhuman. Their well-being is a clear indicator of how civilized society has become. As William Blake observed in his poem, "A dog starved at his Master's Gate predicts the ruin of the State."

## Self-Awareness

Some psychologists, like Professor Gordon Gallup, contend that self-awareness is evident only in humans and the apes. Prof. Gallup published his findings in a psychology journal, claiming that he had the key to objectively determine if animals are self-aware. He came to this fatuous claim after getting various primate species used to seeing themselves in a mirror. When he put a mark on the animals' foreheads, the apes used the mirror to try to remove the spot once they noticed it in the reflection. But the monkeys he tested had no response. So he proposed, on the basis of this simplistic test, that monkeys are not self-aware and only our closest relatives, the apes, possess this trait.

This conclusion is based on reductionistic and anthropocentric (human-centered) thinking and the inflation of a single behavioral response—self-grooming of the spot in the mirror—to be the sole indicator of whether or not an animal is self-conscious. No less a significant indicator of self-awareness in this same context is the *absence* of a behavioral response. Cats, for example, usually ignore their images in a mirror after initially reacting as though the reflection is another cat. They will use the mirror in other ways, seeing and responding to what is going on behind them while they are looking into the mirror.

I do know of some cats and Asian elephants who will use a mirror to wipe a spot off their foreheads, but maybe dogs are less image-conscious—or narcissistic. The anthropocentric thinking of Prof. Gallup and others is problematic. By this I mean interpreting the behavior, sociality, and learning abilities of other animals from the yardstick of how humans would think and act under similar circumstances. Such pseudo-scientific performance determinations, like those of Prof. Gallup, only serve to reinforce the speciesist and racist attitudes of those extreme rationalists, chauvinists, and supremacists who are averse to the ethic of equalitarianism: of giving equal and fair consideration to the rights and interests of all beings, no matter how unintelligent they may seem.

## Animal Altruism

Being altruistic is not an exclusively human virtue. There are many accounts of the incredible feats of heroism by dogs and cats, stories of animals rescuing family members and strangers from fire, drowning, and from dangerous people and other animals. In their altruistic behavior, these animals demonstrate remarkable resourcefulness and insight.

Through altruism and empathy, cats care for each other and thus contribute to the greater good of their community or social group. Altruistic behavior is especially evident when we observe a mother cat grooming her kittens, licking them to stimulate evacuation, and when they are older, patiently training them to follow her and to not play too roughly. Some mother cats are more patient and indulgent than others, which are differences that human mothers share. There are many reported instances of feline altruism and courage—notably cats that have awakened their families to a fire in the house, and cats that have gotten burned themselves rescuing their kittens. Call it instinct if you wish, but it is not blind instinct. Altruism may indeed be inborn, but, like empathy, it is a virtue certainly not exclusive to the human species. To witness such behavior in animals is indeed a humbling experience and helps affirm our biological and spiritual kinship with them.

Cats can help us relax and give untold benefits to the human soul.
Photo Aya Kinoshita

# More about Empathy

If animals were incapable of empathy, of understanding another's emotional state and having feeling for another's distress, then we would find no evidence of altruistic behavior in the animal kingdom. But indeed we do. Ethologists use the terms *care-giving* (epimeletic behavior) and *care-soliciting* (et-epimeletic behavior) to identify those behaviors that underlie the altruism we see in various species, which means that they do have the capacity to empathize.

Studying reactions in mice to other mice given mild noxious stimuli that caused stomachache-like pain, McGill University scientist Jeffrey Mogil and his Pain Genetics laboratory research team concluded that mice have a hardwired form of empathy that they termed *behavioral contagion*. Mice showed more empathy toward a familiar cage-mate than to a strange mouse in distress.[3]

Skeptics continue to dismiss evidence of animals' empathy as anthropomorphic and scientifically unproven, and it disturbs me to read some professional comments on this topic. For example, veterinarian John S. Parker stated that, "Pets can and often react to their owners' distress or discomfort, but that is not to be confused with experiencing the emotion of empathy."[4] Aside from contending that animals "do not have the cognitive capacity to put themselves in our place," he incorrectly sees empathy not as a process or affective state but as an actual emotion, which it is not. Animal ethics philosopher Dr. Bernard E. Rollin's response in this same journal[5] stating that, "there is some very suggestive evidence that at least some animals, such as higher primates and elephants, do [empathize]" begs the question. The evidence from countless instances of empathetic behavior in companion animals is a red flag and not some anthropomorphic red herring, putting us all on notice that animals are far more aware than many people would like or accept, for reasons best known to themselves.

Here are some of the many accounts that people have shared with me about their empathetic animal companions.

Esther Schy from Fresno, California, writes, "When I returned from the Cancer Center following treatment, I was extremely weak and ill. My two Airedale dogs would each take up their positions, like two bookends, one on

---

3. *Science* magazine (June 30, 2006): 1860–1861.
4. Letter in the *Journal of the American Veterinary Medical Association* (June 1, 2006): 1677–1678.
5. Ibid. 1678.

either side of me in bed, and would lie there unmoving for hours, except for their taking turns laying their heads gently where I hurt the most."

One night two years earlier one of her dogs named Robbie "suddenly jumped up in bed next to my husband, almost plastered to his side . . . He normally never did this preferring to sleep on his cushion next to my side of the bed. He kept trembling for one hour, and then went downstairs by himself, which is another action he did not normally do (leaving the bedroom at night). My husband suffered a massive heart attack and died a few minutes later. I believe that Robbie knew that something awful was to be."

Like the Airedales that rested their heads on where their human companion hurt most, M.S.D. from Romeo, Michigan, has a Siamese cat that picked up on her cardiac palpitations that were causing much distress and preventing her from sleeping. "My Chloe came up, got as close as she could, and placed her paw on my left chest over my heart. Within a very short time the palpitations slowed and stopped, allowing me to get a good night's rest."

Amy E. Snyder in Chesapeake, Virginia, was comforted by her Maine coon cat Bonkers, who slept at her side during her ordeal with throat cancer, giving her comfort and constant attention. Taking radiation treatment some one hundred miles away from home, Ms. Snyder was only able to come home on weekends; one weekend she found Bonkers lethargic and looking older. She took him to her veterinarian, and Bonkers was euthanized because he had developed an inoperable cancer, "completely cutting off his windpipe . . . I believe, due to the extreme oddness of similarity to our illness that my cat literally tried to take on my disease. He did get me through all of this."

This anecdote supports my theory of sympathetic resonance, where highly empathetic animals may develop the same or a similar disease that afflicts their loved one. Whether it is deliberate or coincidental, the fact remains that empathizing is not without risk for humans and nonhumans alike.

Patricia Anderson from Osceola, Missouri, had home nursing care after her sojourn in the hospital, and the nurses were moved by her two cats, which she called her Guardian Angel Kitties. "My two gray tabby cats positioned themselves one on each side of me and they would not leave during the procedures the nurses needed to perform. When I was experiencing pain or feeling depressed I just had to reach a few inches and find a warm loving presence. They took turns getting down to eat and using the litter box so I was never alone. One day they both jumped down and began to romp and play in their usual manner. They knew, before I did, that I was healing and getting better."

Many other letters attest to how cats and dogs have helped their human companions cope with depression and other emotional and physical difficulties, especially the loss of a spouse or other close relative. Cary Watson from Clifton Park, New York, writes that, "Without my two dogs' companionship, dealing with the loss of my wife would have been much harder. I can see why many people die soon after losing a spouse. We need love to carry on."

Echoing this sentiment, Barbara K. Joyner of Courtland, Virginia, wrote that, following the untimely death of her husband to be, her adopted cats "make me feel wanted, needed, loved. They bring joy and happiness into my dark, sad existence."

Suffering the loss of her only child from suicide, Patricia Maunu of Sioux Falls, South Dakota, tells me that her bichon frise dog J'aime, "has given me the desire to get out of bed, and on many days given me the will to live!"

These and many other personal stories about how companion animals have helped their human guardians through difficult times and are a constant source of affection and joy help us all appreciate why so many people who were victims of the Katrina hurricane disaster in New Orleans and other communities refused to leave without their animal companions. They are an integral part of the family and of the emotional lives of millions of people, and those who have not experienced the gifts of animal companionship, and the depths of animals' empathy, have missed a golden opportunity to enrich their lives and awaken their appreciation for all creatures great and small.

In the next chapter we will explore how animals can also play a significant role in our spiritual and moral development, as well as contributing in many ways to our emotional well-being.

# 11
# | Animals and Our Spiritual Development |

I have been a veterinarian for over forty years and have worked in animal protection as an animal rights and environmental advocate for over thirty years. My avocation has always been to improve the health and well-being of animals. I have learned a lot about human attitudes toward animals and I am disturbed that there is no unified sensibility. A most important and most difficult type of healing that I see needed in every country and culture—for the benefit of both the human and the nonhuman—is to restore the human-animal bond.

A healthy bond is one that is mutually enhancing. Two essential ingredients are required. First, a better knowledge of ecological and ethological science: the objective understanding of animals' environmental needs, behavior, communication, and emotions. Second, a better subjective understanding of animals as sentient, feeling beings. This requires establishing an empathic connection with animals. Empathy is the ability to share or experience another's feelings.

Empathy is the emotional bridge that connects us with others. Scientific knowledge about animals does not establish a bond; empathy does. Scientific knowledge helps us better understand animals' needs, and gives us a greater clarity in interpreting their behavior, intentions, and emotional state. But it does not help us feel for animals. Feeling for animals comes from the heart—the core of our being more ancient and more sapient than all the factual information our intellects hold about the nature of ourselves and the sentient world.

## Compassion and the Origin of Ethics

As Pascal observed, "The heart knows what reason knows not," meaning that wisdom comes not from the intellect alone, but also from our emotions. It is from our feeling for others that our moral sensibility and ethics arise.

Ethics and moral codes cannot make us feel for others. Our feeling for others comes through caring concern or mindfulness. The antithesis is to not be *mindful* of others, to be ignorant because of selfish preoccupation. Selfishness is thus the root of ignorance and makes empathy impossible.

Psychiatrist R. D. Laing discusses the debasement of our humanity, the true nature of human beings, when we harm animals:

> *A woman grinds stuff down a goose's neck through a funnel. Is this a description of cruelty to an animal? She disclaims any motivation or intention of cruelty. If we were to describe this scene "objectively" we would only be denuding it of what is "objectively" or, better, onto-logically present in the situation. Every description presupposes our ontological premises as to the nature (being) of man, of animals, and of the relationship between them. If an animal is debased to a manu-factured piece of produce, a sort of biochemical complex—so that its flesh and organs are simply material that has a certain texture in the mouth (soft, tender, tough), a taste, perhaps a smell—then to describe* the animal *positively in those terms is to debase oneself by debasing being itself. A positive description is not "neutral" or "objective." In the case of geese-as raw-material for pate, one can only give a nega-tive description if the description is to remain underpinned by a valid ontology. That is to say, the description moves in the light of what this activity is a brutalization of, a debasement of, a desecration of: namely, the true nature of human beings and of animals.*[1]

Poet Gary Snyder wrote, "All creatures are equal actors in the divine drama of awakening. The spontaneous awakening of compassion for others instantly starts one on the path of ecological ethics, as well as on the path to enlightenment."[2]

1. R. D. Laing, *The Politics of Experience and the Birds of Paradise* (New York: Penguin Books, 1967).

2. Gary Snyder, *A Place in Space: Ethics, Aesthetics and Watersheds* (Washington, D.C.: Counterpoint, 1995).

But what awakens concern for others and ignites the passion needed to sustain compassionate action? Perhaps the awakening is in our sudden recognition of the universal nature of selfhood and of the capacity of all living beings to be harmed and to suffer. Other beings reflect something of our own selfhood in their fears and hopes, pains and pleasures. Through empathy we experience what others feel.

One reader of my syndicated newspaper column "Animal Doctor," after he had caught but not killed a mouse in a spring trap in his kitchen, wrote: "I hastened to take it outside and as I was about to fling it to the ground, I just then caught a glimpse of the mouse's face. I swear, I could *see* the pain and fright in the mouse's eyes."

The late biologist and noted author Loren Eiseley wrote, "We do not find ourselves until we see ourselves in the eyes of those who are other than human."[3]

When we more fully empathize with each other and with other animals, we achieve a higher resonance of being and a greater refinement and clarity of spirit. We discover the value of other beings to our own spiritual growth and self-realization. Animals as *kin* extend our sense of community. Animals as *others* deepen our understanding of evolutionary and adaptative processes. Animals as companions enrich our lives, and their healing powers have been shown to benefit us physically and psychologically. Animals as totems and icons have served humanity for millennia as revelatory symbols of divine creation—as mystical, metaphorical, and mythological images have awakened our imaginations, inventiveness, and religious sensibilities. Animals can take us outside of ourselves into other realms of being and consciousness. We return from the journeying into the empathosphere like the shamans, seers, and healers before us, a little more enlightened and inspired by our communion with other animals, some of whom are our allies, familiars, significant others, and also our teachers and healers.

Animals as objects and subjects serve our many needs and appetites, often reflecting in their plight the dark side of human nature. They teach us that when we demean them, we demean our own humanity.

Through reason we acquire objective knowledge. Through empathy we acquire subjective knowledge. When these two modes of understanding are integrated, then we have wisdom. When we understand animal behavior

---

3. Loren C. Eiseley, *All the Strange Hours* (New York: Charles Scribner's Sons, 1975).

and empathize with animals, allowing compassion to be the ethical compass of reason and action, then animals help us grow in wisdom—and help us evolve into a more humane species, worthy of the name *Homo sapiens*.

As the "unfinished animal," I see my species on the threshold of an evolutionary transformation into a panempathic being, having feeling for all life, provided we can divest ourselves of the anthropocentric (human-centered) worldview of human superiority and dominion. Humility is a prerequisite for the birth of our empathetic powers and faculties, as we replace arrogance and fear with compassion and love. Compassion is the ethical compass for our power of reason, without which the human intellect cannot function optimally, and will continue to cause more harm than good.

In the presence of another animal we can experience the oneness of existence, of being alive, sentient, vulnerable, and mortal. Yet in the same gaze or breath we can experience also the profound otherness of a different species across a vast distance not in the dimensions of time or space but of consciousness. The animal looks into our eyes, and there is a moment of recognition—of affirmation, perhaps. Then the gaze seems to pass through and beyond us. At this moment the animal mind becomes incomprehensible, something profoundly other, part of a deeper mystery that this sentient universe unfolds and enfolds. What is revealed in one moment of communion is veiled in the next. The ancient Indian text, the Brihad-Aranyaka Upanishad, puts it this way: "The immortal is veiled by the real. The Spirit of Life is the immortal. Name and form are real, and by them the Spirit is veiled."

So in the reality of an animal's presence, the immortal Spirit of Life is unveiled when we enter into communion. The paradoxes of oneness and otherness, of unity and diversity, and of duality and nonduality, are transcended. Such is the magical, transformative power of animals on the human psyche when we allow ourselves to be open to their presence and when our being-in-consciousness resonates with theirs. However fleeting the moment and forever enduring the memory of such sympathetic resonance with another living soul, we know that what was unveiled was indeed the Spirit of Life immortal embodied in the archetypal form that we know by name as eagle, whale, or wolf.

Some spiritual teachers, mystics, and poets advise that when we see the universal in the particular, and the particular in the universal, then we will enter the Kingdom of Heaven, or what others call the enlightened state of Samadhi or Nirvana. Those who believe or assert that this is a state of blissful liberation from sufferings and tribulations of the sentient world are wrong if they speak only of the liberation of the particular (one's self) and not of universal liberation. After personal enlightenment, then what? There is no lasting enlightenment when empathy with other sentient beings is severed and one is no longer connected to the unified field of being and part of the boundless circle of compassion.

Those who are enlightened about the universal nature of self ask themselves how they can sustain their own lives with the least injury to the lives around them. How can they help prevent and alleviate the suffering of the world in which they live? Personal liberation and animal liberation from cruel human exploitation are part of the same spiritual activism that arises out of radical compassion that is boundless in its empathic embrace of all life.

The ultimate value and significance of animals to us, other than their roles in the wild contributing to the health and functional integrity of aquatic and terrestrial ecosystems, is to facilitate the evolution of our humanity, our spiritual development. When we empathize with them, we become more humane and aware. In the process, they ennoble us in spirit as our respect and devotion for them grows from generation to generation, and from age to age.

## Animal-Insensitivity Syndrome: A Cognitive and Affective Developmental Disorder

Animal-Insensitivity Syndrome, an impaired sensitivity toward animals, is a disorder that I see as part of a larger problem of insensitivity and indifference to the Earth.

Ethical blindness that comes from a lack of empathy with other living systems and beings is linked to a lack of respect and understanding, so that when we harm animals and the Earth, we harm ourselves—especially in our production of food and fiber, and indirectly in our dietary choices, consumer habits, and lifestyles.

We harm animals by destroying their natural habitats, and in making them suffer so that we may find new and profitable ways to cure our many diseases. The actual prevention of disease is in another domain based on an entirely different currency from what is still the norm in these sickening times. The currency of unbridled exploitation and destruction of natural resources and ecosystems, and the wholesale commercial exploitation of

The child is father to the man. A little boy enjoys communion with kittens, and as an adult with a rescued cat.
Photos M. W. Fox

animals, cannot continue because it is not sustainable. One of the greatest sicknesses is the proliferation of factory livestock farms—the intensive confinement systems that are stressful to the animals, promote disease, are environmentally damaging, and also put consumers at risk.

These animal concentration camps of the meat, dairy, and poultry industries will only be phased out when there is greater consumer demand for organic, humane, and ecologically sustainable animal produce for human and companion animal consumption.

So I was heartened to see that author Richard Louv has written a book entitled *Last Child in the Woods: Saving Our Children from Nature-Deficit Disorder.*[4] This is the flip side of the coin that shows: heads, nature; tails,

---

4. Richard Louv, *Last Child in the Woods: Saving Our Children from Nature-Deficit Disorder* (Chapel Hill, N.C.: Algonquin Books, 2006).

animals. Both are in our hands, for better or for worse. Louv contends that in modern industrial consumer society, children are being raised and educated without meaningful contact with or any understanding of the natural world, and that this will be bad both for them and for the future of the Earth. This has caught the attention of educators and parents around the world. Long before this important book was published, my friend Jane Goodall, the famed chimpanzee biologist, had recognized this issue and initiated her "Roots and Shoots" program in schools in many countries to educate children about nature, ecology, and the inherent value of wild plants and creatures.

In the common currency of compassion and respect, our transactions and relationships with each other, with other animals, and with the Earth or natural world are framed within the Golden Rule. This rule, embraced by all world religions, is to treat others as we would have them treat us—and in the Golden Rule, gold alone does not rule. This currency includes ancient coins including: wisdom as *altruism*, that is, enlightened selfishness; *ahimsa*, Sanskrit for not harming in any way; and *karma*, having prescience and understanding that what goes around, comes around. All our choices and actions have consequences.

Sustainable rates of exchange are based on mutual aid, a point emphasized by Russian Prince Peter Kropotkin (circa 1900), who envisioned the ideal human community like a functioning ecosystem of interdependent, democratically integrated individuals and species creating mutually enhancing, symbiotic, micro- and macro-communities—a concept he discovered in his studies of the co-evolved flora and fauna of the vast wild steppes of his native land.

Nature-Deficit Disorder leads ultimately to regarding and treating the living Earth as a nonliving resource, just as the Animal-Insensitivity Syndrome can lead to animals being treated as mere objects without feeling. Insensitive, indifferent, and cruel contact and experiences with animals during the early years (probably a critical sensitization/desensitization developmental stage between eighteen and thirty-six months of age) can mean a poorly developed and extremely self-limiting capacity to empathize with others; to be able to recognize, anticipate, experience, and share other's feelings; and to express and deeply consider one's own. Adult denial and ethical blindness are rooted in early childhood conditioning and desensitization.

In some countries where I have worked with my wife Deanna Krantz, such as India, we have both witnessed how people simply turn a blind eye to the suffering animal and the polluted stream because they themselves are struggling to survive. Individuals who feel helpless become resigned fatalists, or they are either too lazy, busy, desensitized, or blind to lift a finger to try to make a difference. In some contexts intervention to help a suffering animal, or to stop a stream from being poisoned by a tannery or a slaughterhouse, could mean death threats and violence.

Observing another's suffering, and being unable to do anything to help, leads to learned helplessness. Seeing other's suffering, and being indifferent about it, is the next step toward the total disconnect of empathy, called *bystander apathy*. The next step is to observe and derive vicarious pleasure in witnessing another's plight. This is but one small step away from deliberate torture and calculated cruelty, either perpetrated alone or in participation with others—in the name of entertainment, politics, sport, quasi-religious or cult ritual, and, as some see it, experimental vivisection.

Why does it matter if animals must be made to suffer and die and the natural environment is obliterated, as long as human needs and wants are satisfied? For many people it obviously does not matter—even when their values and actions harm those who do care and feel it *does* matter and that it is morally wrong to harm and kill animals and destroy the natural environment. The ethics of compassion and *ahimsa* mandate that we find the least harmful ways to satisfy our basic needs and relinquish those wants, appetites, and desires that cause more harm than good. Such renunciation is seen by some as the only hope for humanity and for our sanity: to live simply so that others may simply live.

Animal suffering matters because it is a matter of conscience. Deliberate cruelty toward animals and acceptance and indifference toward their plight is unconscionable: this is a *zoopathic* state of mind. This parallels the behavior—and cognitive and affective impairment—of the sociopath and of the *ecopath* who has no twinge of conscience over the destruction of the natural environment. Where there is a lack of empathy, of feeling for others, there can be neither concern nor conscience.

Becoming desensitized to animal suffering and then treating animals as mere things, as objects devoid of sentience, is part of the same currency as treating fellow humans as objects. Such dehumanization, coupled with demonization, can lead to genocide, and more commonly to *speciecide*. This is the annihilation of animal species and their communities that are perceived as a threat. Our attitudes toward other animals—our degree of ethical concern and moral consideration—can mirror our regard for each other, for better or for worse. When collectively our hearts and minds are open to the tragedy of reality, and we really see and feel all that is going on around us and empathize fully with the suffering of others, times will begin to change for the better.

The antidotes are many. Those in Richard Louv's book should be coupled with meaningful contact with companion and other animals, with parental supervision and humane instruction to foster respect, self-restraint, gentleness, patient observation, and understanding.

 A child's sense of wonder, if it is nurtured and not crushed or left to wither, blossoms into the adult sense of the sacred: an ethical sensibility of respect for the sanctity of all life.

A child's sense of curiosity leads to natural science and instrumental knowledge. Combined with a sense of wonder, curiosity leads to imagination and creativity, while the sense of the sacred is the foundation for an ethical and just society, and for empathetic, caring, and fulfilling relationships, both human and nonhuman. This empathy-based bioethical[5] and moral sensibility that gives equally fair consideration to all members of the biotic community—human and nonhuman, plant and animal—is an ideal that may yet become a reality, provided the potential for such development is properly nurtured and reinforced by example in early childhood.

Like our physical health, our mental health and Earth health are deeply interconnected, and for us to be well and whole in body, mind, and spirit, our connections with animals and the Earth must be properly established in early childhood in order to prevent the harmful consequences of Nature-Deficit Disorder and Animal-Insensitivity Syndrome. Our collective inertia over doing anything constructive to address such critical issues as human population growth, overconsumption, pollution, global warming, and the plight of animals domestic and wild will then become something of the past. Then initiatives—local and international—to promote planetary CPR (conservation, preservation, and restoration) and to promote the humane treatment of animals will become a reality because, in the final analysis, it is in our best interests to do so.

---

5. See Michael W. Fox, *Bringing Life to Ethics: Global Bioethics for a Humane Society* (Albany, N.Y.: State University of New York Press, 2001).

When we harm animals and the Earth, we harm ourselves, and the generations to come will all suffer the consequences of our actions and inaction. As the Iroquois Confederacy advised, the good of the life community mandates that, "we think seven generations ahead, and seven generations back." This translates into the bioethical consideration of consequences and in practice means that those who do not learn from the mistakes of their ancestors shall live only to repeat them.

## Animals and the Politics of Compassion

It is a bioethical imperative for us humans to develop the wisdom, which in other animals is primarily instinctual, that enables us to live in harmony with all beings. This means not causing harm to others and therefore to oneself, because the self is realized fully through the love it *gives* rather than seeks only for itself. As the Buddha advised, the only true religion is *maitri* (loving kindness or benevolence) toward all creatures. He also taught that the end of suffering is in suffering itself, meaning that when we embrace the suffering of others through empathy, the communion of suffering becomes a communal response to the plight of others, human and nonhuman, who are then helped because the pain of one is the pain of all.

There is a wonderful story of a Buddhist monk who did not become enlightened until he put compassion into action. In his book *Universal Compassion: Transforming Your Life Through Love and Compassion*, Geshe Kelsang Gyatso writes:

> *Asanga, a Great Buddhist Master who lived in India in the fifth century AD, meditated in an isolated mountain cave in order to gain a vision of Buddha Maitreya. After twelve years he still had not succeeded and, feeling discouraged, abandoned his retreat. On his way down the mountain he came across an old dog lying in the middle of the path. Its body was covered in maggot-infested sores, and it seemed close to death. This sight induced within Asanga an overwhelming feeling of compassion for all living beings trapped within samsara. As he was painstakingly removing the maggots from the dying dog, it suddenly transformed into Buddha Maitreya himself. Maitreya explained that he had been with Asanga since the beginning of his retreat, but, due to the impurities in Asanga's mind, Asanga had not*

*been able to see him. It was Asanga's extraordinary compassion that had finally purified the karmic obstructions preventing him from seeing Maitreya.*[6]

That which purports to be spiritual must translate into compassionate action, loving relationships, and understanding. If it does not, then it is not spiritual because the essence of spirituality is to live ethically "in a sacred way," as the late Sioux medicine man Black Elk advised. The ethic of what Albert Schweitzer called reverence for life is the key directive. In our relationships with other animals, we have a duty to respect their fundamental entitlements of being, such as the freedom to be themselves: for birds to fly and not be in cages; for whales and dolphins to swim the oceans and not be in aquariums; for pigs and cows in factory sheds to run in the fields; and for dogs to be dogs in off-leash parks with each other. All creatures under our control and stewardship have a right to freedom from pain and fear arising from how we humans so often treat them. Some moral philosophers call these entitlements *animal rights*.

I received the following question from a reader of my Web site (www. doctormwfox.org):

**Question:** A friend of mine says dogs don't have souls. Even so, he says they should be treated humanely because otherwise it's a sign of bad character. God gave us dominion over them, to use as we choose, but it should be kindly use. What is your opinion?

**Answer:** *Kindly use* can be a slippery slope. I interpret dominion as loving kindness, as in God's dominion over us. If we are created in God's image, then we should treat dogs and all fellow creatures as we would like God to treat us.

As the late and controversial Pope John Paul II has said, all creatures, like humans, are enspirited "with the same breath of Creation." Being part of the same Creation, we should therefore give animals equally fair consideration. In my recent book *Bringing Life to Ethics*, I call this cardinal bioethical principle *equalitarianism*.

---

6. See Geshe Kelsang Gyatso, *Universal Compassion: Transforming Your Life Through Love and Compassion* (Glen Spey, N.Y.).

I don't believe animals *have* souls. Like humans and plants, they *are* living souls. In my metaphysics of what I call *biospiritual realism*, the spirit is not in the body. The body is in the spirit. Through this primal and sacred duality, souls are born to experience life in different forms. Many people embrace this spirituality of the oneness, individual sanctity, and interdependence of all beings—but many don't. That is why some find it morally repugnant to put human genes into pigs and to conceive a biotechnology industry dedicated to cloning organ-donor pigs, endangered species, and people's pets and children. But many don't, and I believe that these unfeeling and therefore unawakened people are not fully compassionate humans. They are responsible, I believe, for what Charles Darwin implied in the title of his book *The Descent of Man*.

In other words, the human species has become an ethically degenerate, morally challenged, and empathically impaired species that is a blight, a parasitic infestation, on Mother Earth. The late Prof. Konrad Lorenz, one of the founding father-scientists of ethology and an avowed pantheist who believed that all animals were as much a part of the sacred as we are, wrote: "Far from seeing in man the irrevocable and unsurpassable image of God, I assert—that the long-sought missing link between animals and the really humane being is us!"

In his seminal book *The Eternal Treblinka: Our Treatment of the Animals and the Holocaust*, Charles Patterson writes: "Throughout the history of our ascent to dominance as the master species, our victimization of animals has served as the model and the foundation for our victimization of each other. The study of human history reveals the pattern: First, human beings exploit and slaughter animals; then they treat other people like animals and do the same to them."[7]

Columbia River Tribes activist Ted Strong echoes these sentiments, stating, "If this nation [the United States] has a long way to go before all our people are treated equally regardless to race, religion or national origin, it has even farther to go before achieving anything that remotely resembles equal treatment for other creatures who called this land home before humans ever set foot on it."

The misinterpretation of Darwin's theory of the *survival of the fittest* links his findings with the inverted morality of *might makes right*.

---

7. Charles Patterson, *The Eternal Treblinka: Our Treatment of the Animals and the Holocaust* (New York: Lantern Books, 2002).

This misunderstanding violates the cardinal principle of Natural Law, namely, mutual cooperation. The notion of some God-given—and socially, politically, and religiously condoned power over others—needs to be dispelled. Charles Darwin's use of this term was in reference to environmental fitness or adaptability, not power and competition. He recognized how important cooperation was within and between species. Survival through fitness is quite different from survival through power and control over others. Yet his theory of evolution through natural selection and survival of the fittest was seized upon by a very class-conscious English society that condoned industrialism's power over nature and colonialism's imperial power over and enslavement of other cultures and nation-states. Darwin's theory was twisted out of context to give scientific credence to a survival-of-the-fittest mentality (read, superior) that sanctified competitive individualism that was encouraged in children from kindergarten on. Coupled with the religious (Judeo-Christian) belief in man's superiority over lesser beings and nature (that were made by God for man's use), this mentality has made us the least-fit species on Earth because of the harms done to others, to the Earth, and ultimately to ourselves.

A spirituality that does not bring love to life is like a philosophy that does not bring ethics to life—like a religion that sees no spirituality in nature, and a spirituality that has no immediate social and political relevance. The kind of ethics that we need to bring to life, like equalitarianism and reverential respect for all living beings, are the intellectual fruits of reason that are ripened by emotion—especially our empathic and intuitive sympathies, passion for justice, and our power of love rather than the love of power. If our love for fellow creatures and for nature has no social and political relevance, then it is not true love but selfish attachment (for various reasons such as emotional, esthetic, and pecuniary). It would be pedantic and preachy for me to document why: But why isn't everyone who lives with a dog or a cat concerned about the welfare of all dogs and cats and actively supporting at least one reputable animal protection organization or local animal shelter? Why isn't everyone who eats animals concerned about how factory-farmed food animals are generally raised for human consumption and how a meat-based diet impacts wildlife habitat and biodiversity, and are then moved to do something about it? Any dominant species that chooses not to think or to empathize deserves to become sick, to be afflicted by a host of the so-called diseases of civilization—suffering the consequences of their own actions and inaction and, for the greater good, become quickly extinct.

To love the universal in the particular, and the particular in the universal, is to embrace, nurture, and defend the freedom for the particular to be, and for the universal to become. When we protect all our animal relations (or biological ancestors) with the same passion and commitment as we might protect our own kith and kin, then the human will become a humane and ethical animal. Some call this self-realization, while others call it human evolution.

In other words, such love respects, protects, and nurtures the freedom of all living beings, which leads to animal and human rights and liberation. It also seeks to secure the integrity and sanctity of creation and universal becoming, which is the spirit of the deep ecology, Earth First!, animal rights, and holistic health movements—which are seen by some as anti-establishment, terrorist organizations that are a threat to both the economy and to national security!

A deeper understanding of the primordial, coevolved, empathic relationships between bee, flower, and meadow, and deer, wolf, and forest, means a deeper appreciation and respect for the creative dimensions of love where life gives to life to sustain a greater whole—the biotic community. Where in relation to this ecology and spirit of mutual cooperation that is self-sustaining and self-affirming do we fit in, with faster and bigger gas-guzzling vehicles and other insatiable, media-hyped appetites that are laying waste to the environment and pushing countless animal and plant species into extinction, as the others we exploit suffer the consequences of our greed, ignorance, and need?

These are the ethical, economic, environmental, spiritual, political, and metaphysical concerns of the deep ecology, Earth First!, and human and animal rights movements. Opponents seek to discredit them in order to protect their own vested interests by preserving the status quo. Opponents call such concerns anti-progress, and anti-human. But it is an Earth- and life-centered love and respect that provide the ethical basis for environmental and public health, for conservation agriculture and organic farming practices, and for a sustainable economy that places the human within rather than above or outside the Earth community. Such spirituality sees the human as part of the creative matrix of self-organizing, intelligent, and transformative processes that some call God, or nature, or sacred creation. When we submit to and learn to live in harmony with what Sioux medicine man Black Elk called "the Sacred Power of the world as it lives and moves," all will be well. The way of harmony is therefore the way of loving kindness and selfless service.

# 12
## | Feeling for Animals and Healing the Bond |

Millions of people enjoy the company of their cats because they give so much pleasure, expressing affection, playfulness, concern, interest, joyful greeting, and grateful satisfaction. These and other emotive behaviors we do not imagine. They are as real as the feelings they evoke in us. The subjective state or feelings of humans, cats, dogs, and other sentient beings when expressing various emotions must be similar, since they evoke similar affective responses.

The gulf that separates us from other animals is not our intelligence but our disbelief: our inability to accept their similar sentience and their emotional affinity with us. We should not demean their affinity as pathological dependence, or as the subversion of unconscious instincts by domestication. It is too narrow a view to believe that adult cats regard their "owners" simply as parent-providers and that dogs would sooner be with other dogs but are bonded to their "masters" as substitute pack leaders.

As our understanding of the *ethos* of cats, dogs, and other creatures— both wild and tame—deepens, so must our understanding of the human-animal bond and of people's attitudes toward animals in various cultures, as well as psychohistorically. Otherwise our knowledge will be incomplete, and we will fail to fully comprehend much beyond self-serving utilitarian horizons. The "objective" language of the psychobiological sciences are not much help, tending to limit our horizons with mechanistic terminology. It is recognized more widely now that animals often display behavior patterns that communicate to us (empathetically and via their emotion-evoking distress calls, expressions, and postures) a whole range of subjective states, such as separation anxiety, despair, grief, terror, as well as relief, playful good humor, satisfaction, concern, and even empathy, compassion, and altruism. But the semanticists of scientism have no objective synonyms for these subjective states. And so it is considered unscientific to allude to such states, but in truth we have no other words that do not sound anthropomorphic. Alternative, mechano-morphic terms like *approach*, *avoidance*, *satiation*, and *consummation* are unacceptably simplistic and give no intimation as to emotional or motivational state.

Some of these very human emotions I felt and witnessed in a Kashmir, India, village some years ago when I saw a male pariah dog licking the sores and snapping at the flies around his mate's infected face. Both were emaciated, and he let the old bitch feed before him after I had thrown them some food. Such empathy and altruistic behavior is common to many animal species, not only to the human species. This ethologically documented evidence of deliberate care-giving behavior to others that are not simply immature offspring is proof of a highly developed empathic sensibility in animals. Some investigators have claimed such sensibility, albeit rudimentary, in trees and other plants, as well as in insects and other invertebrate and protoplasmic life forms. Be that as it may, when we begin to appreciate the significance of the sentient nature of other animals and of the universe itself, our attitude toward and treatment of animals changes radically toward what might best be termed *panempathic communion*.

But when we do not live close to animals in their natural environs and see them only in the altered states of domestication and captive confinement, we can become alienated and unfamiliar with their ways, their ethos. So we deny them feelings, awareness, even sentience as well as intelligence, and in the process of denial we mechano-morphize them, seeing and even treating them as unfeeling machines. Or we anthropomorphize them, unable to see them as they really are through the cloudy lens of sentimentality and objectification.

People who really see animals for the first time as fellow sentient beings become humane because they experience in other animals the kinship of sentience that includes the capacity to suffer and to experience pleasure. This is not some deluded anthropomorphic projection, but a fact of rational human experience and responsiveness and the consequence of feral vision.

I recall reading of an entire change in attitude of an Australian government dingo exterminator. He was hiding in the bush near a spring where tracks indicated to him that dingos came to drink at dusk, ready to shoot whatever wild dogs came. A pair came to drink, and he killed one, the dingo who was slower and seemed to be led by its mate to the water hole. The government agent was shocked to find that the dingo he had shot was blind, had been blind for a long time, and was in excellent physical condition. Without the empathy and concern of his mate, this old blind dingo would have long perished. Realizing how caring dingoes can be for each other in the Outback, the man never killed another dingo from that day on.

There are instances of wild animals seeking human assistance, like a bobcat that crawled out from cover to lie across the skis of a cross-country skier and allowed itself to be wrapped in a jacket and taken to a veterinary hospital to have festering porcupine quills removed. And what of the behavior of a rehabilitated orphan bobcat who brought her cubs, born in the wild, back to show the woman who had rescued and healed her?

There are also accounts of animals who had been treated at a veterinary hospital taking an injured stray animal back to the same hospital for treatment. These and other eye-opening anecdotes reveal a degree of intelligence or sapience as well as sentience. They move us to establish a new covenant of compassion with the entire sentient world. And until we are so moved and our empathy and feral vision close the distance between us and fellow creatures, the possibility of a truly humane society will forever elude us. We will also fall short of realizing our full human potential to be humane. To be humane means treating all living beings—human, plant, and animal—with compassion and reverential respect, including their environments and communities. It is not simply a name for animal welfare or protection, but a state of being in awareness that is the hallmark of a truly civilized society.

## Healings Shared

The many benefits that animals provide people are being recognized by mainstream healing fields that involve the use of animals as pet-facilitated cotherapy, such as animal-assisted living for the physically handicapped. Many have realized the benefits of *hippotherapy*—therapeutic horse riding. Even simply observing wildlife and enjoying a well-maintained aquarium of tropical fish can be therapeutic in many ways to adolescents and nursing home residents respectively.

A cat at school is a catalyst for humane education, and grooming a cat helps instill compassion and nurturing at an early age.
Foxfiles

These new fields I call *zootherapy* are complemented by *anthrozootherapy*, where humans benefit animals in many ways (and often themselves in the process): for example, by being a volunteer dog walker or cat groomer at the local animal shelter, or by helping raise infant wildlife at a rehabilitation and wildlife conservation center or zoo. Many good people take their animal companions to nursing homes, to children's hospitals, and to correctional facilities (some of which have animals already living with and being cared for by inmates).

Taking your own animal companion to a local grade school—or taking a calm, friendly, healthy one from the animal shelter—and talking about what makes animals happy and about animal care, respect, and understanding is part of such therapy.

By so helping promote compassion and humane values to children, the human-nonhuman animal bond has a chance of being formed at an early age during what I believe to be a critical period in the emotional and ethical development of our species. When that bond is either not established, or is malformed, the empathic, emotional sensitivities of adulthood are limited, and thus also our emotional intelligence and ethical sensibility.

The emotional intelligence of various companion animals was recently documented and celebrated in *People* magazine in an article entitled "Superpets!"[1] For example, Zion, a yellow Labrador dog, was playing fetch with a stick thrown into the river when he rescued a boy he did not know. Eight-year-old Ryan Rambo was being swept down Colorado's swift Roaring Fork River, and Zion let the boy grab his collar and swam with the boy holding on for dear life to the shore.

Then there's Daisy, a 150-pound potbellied pig living in Las Vegas who saved one of her human family members, seven-year-old Jordan Jones, from being attacked by a neighbor's free roaming dog. Daisy sustained serious injuries but eventually made a full recovery.

Gary Rosheisen of Columbus, Ohio, who is severely handicapped, tried to teach his orange tabby cat Tommy how to paw the speed dial on his speakerphone to call 911 in case he needed help. But he stopped, because he thought Tommy might start making unnecessary calls. But one day he did collapse on the floor, needed help, and could not reach the phone. But help arrived quickly in response to a 911 call from his number. The only other being in the home was Tommy, who must have made the call.

Super-bunny Robin, living in Illinois, had been purchased by the Murphy family at a garage sale. Ten days later the little rabbit awakened Ed Murphy by making a ruckus in her cage, and continued when he went back to bed. He then discovered that his pregnant wife Darcy, who had gestational diabetes, was barely breathing and he called 911. Darcy is convinced that she and her baby would not be alive today if Robin had not known something was wrong.

Steve Werner of Brentwood, Missouri, is grateful for his old golden retriever Wrigley. She started repeatedly sniffing into his right ear; after seeing a TV report about dogs being able to detect cancer in humans, he decided to get a medical checkup. An MRI scan revealed a tumor the size of a Ping-Pong ball near his inner ear that was surgically removed. It was fortunately benign, but if not detected early, thanks to his dog, it could have caused serious problems.

1. "Superpets!" *People* magazine (September 2006): 168–176.

As Hippocrates, one of the fathers of modern medicine, advised, "Physician, first heal thyself." We see in the human-nonhuman animal bond such wisdom affirmed where the human is healed, made to feel more whole and happy, and in the process the animal feels good, enjoying life by giving and receiving affection. Well-being and welfare depend on physical and mental health, and in the mutually enhancing symbiosis between people and their animal companions or therapy animals, a kind of reciprocal healing may occur in the feel-good interaction between two sentient beings.

## Healing Animals, Animal Healing

I quickly learned as a veterinary student some forty years ago—working on various farms and zoos and seeing large and small animals, wild and domesticated—that many of the health and welfare problems of animals were caused by:

1. How they were treated

2. How they had been selectively bred to have various desired traits and acquired genetic abnormalities

3. How they had been raised

4. The kind of environment and relationships they had during early development with their own kind and with their human caretakers

5. The kind and quality of food they were fed

6. The living conditions they were provided to satisfy their basic physical and psychological needs

I came to realize that the healing of animals and their well-being entailed healing the human-animal bond. This bond was in urgent need of repair because so much of the sickness and suffering of animals was due to how people perceived them and treated them. It also dawned on me that if people were not in a right, ethical, and equitable relationship (or in their right minds, so to speak) with fellow creatures, then the same probably held true for their relationships with their own kind and with the environment.

So the key to preventing much human suffering and sickness was probably the same as for animals: the key was one of relationship and mindfulness, of compassion and empathy, of understanding and caring attentiveness.

## The Ethics of Using Animals in Biomedical Research

This realization was hammered home many times when I visited various university laboratory animal research facilities and saw the small cages and pens in which cats, dogs, rhesus monkeys, and other creatures were incarcerated. What shocked me was seeing arboreal and highly social primates being housed for years alone in metal cages measuring three feet by four feet; and seeing cats and dogs that were formerly family pets being kept in impoverished environments in which they were clearly disoriented, deprived, and distressed.

Part of William Hogarth's eighteenth-century engraving documenting animal cruelty.

I investigated and reported how these seriously inadequate conditions for laboratory animals created sick, stressed, and abnormal animals, making them dubious subjects for scientific study. Research findings from these animals had little medical relevance, except perhaps to other creatures kept under comparable intolerable conditions.

While I saw—as a veterinarian and biomedical scientist—the status quo of laboratory animal care and use as being ethically and scientifically untenable, most of the U.S. biomedical research establishment rejected my findings and continued to defend all vivisection unconditionally.

The biomedical research industry has so much political, economic, and public power and influence that it can change truth and revise history. With a few phone calls and threatening letters, they demonstrated this power by making the publishers of the *Encyclopedia Britannica* alter my contribution to the 1991 revised edition in subsequent editions, deleting all my references to examples of cruel experimentation and to animal rights in the Dog entry, as detailed below. (I also did the Cat entry for this edition of the encyclopedia, but that was no problem since I was not allowed to put in anything about some of the horrendous experiments done on cats—especially popular in brain research and spinal trauma studies.)

*Enclycopedic Knowledge or Convenient Knowledge?*
It was no coincidence, since animals were my teachers as a child, that as an adult I should teach others about the ways and whys of animals. My formal education enabled me to acquire the appropriate scientific terminology to describe animals' behavior, and my lessons from the animals helped me interpret their behavior, develop new theories, and challenge some established and establishment views. Indeed one of the many gifts of the animals was my gaining international recognition as an animal behavior expert, which opened many doors, including the opportunity to contribute to the esteemed *Encyclopedia Britannica*.

I had a golden opportunity to help atone for my own sins of omission and commission when I was given the task of writing updated revisions for the Cat and Dog entries to go into the 1991 edition of the encyclopedia.

I decided to include the following factually accurate paragraph on the use of dogs in biomedical research, because I felt I owed something to my best friend and teacher, the dog:

> *Another common use of dogs, especially purpose-bred beagles, is in biomedical research. Such use, which often entails much suffering, has been questioned for its scientific validity and medical relevance to human health problems. For example, beagles and other animals have been forced to inhale tobacco smoke for days and have been used to test household chemicals such as bleach and drain cleaner. In addition, dogs have been used to test the effects of various military weapons and radiation.*

Soon after the new edition of the encyclopedia was published, the editorial offices were deluged with hundreds of letters from individual scientists, from their prestigious associations like The American Society for Pharmacology and Experimental Therapeutics, the Society for Neurosciences, American Physiological Society and Federation of Biological Scientists, and also from the general public who belonged to pro-animal experimentation and exploitation organizations such as Putting People First and The Incurably Ill for Animal Research.

The biomedical scientists contended that my statement was biased and unbalanced because it didn't emphasize how experiments on dogs have helped people suffering from diabetes, or those needing a coronary bypass or a new hip or kidney. My response was that the entry had nothing to do with the benefits of using dogs in biomedical research.

I offered to change the "offensive" statement as follows:

> *Another common use of dogs, especially purpose-bred beagles, is in biomedical research. Such use, which often entails much suffering, has provided some scientific and medical insights, but is now being questioned on ethical grounds and for its scientific validity and medical relevance. For example, dogs and other animals have been forced to inhale tobacco smoke for days; used to safety test household chemicals, such as bleach and drain cleaner; and used to evaluate the effects of various military weapons and radiation.*

But the editor of the encyclopedia would have none of it. He insisted on substituting the following statement in the next edition to read:

> *Another common use of dogs, especially purpose-bred beagles, is in biomedical research, a use that dates at least to the 17th century.*

I told him that this was totally unacceptable unless he added that anesthetics weren't developed until the nineteenth century. Otherwise, in simply stating that dogs have been used in biomedical research since the seventeenth century, he would be ignoring the fact that this means two hundred years of vivisecting dogs (and countless other creatures) without any anesthetic, which is illegal today. Also such a statement implies some kind of ethical acceptance based on historical precedence. The fact that dogs have been experimented upon by scientists for three hundred years is no basis for justifying, or not even questioning, the continuation of animal experimentation and the continued suffering of animals.

Clearly, I had offended the academic establishment, yet my scientific credentials and long list of scientific and popular articles and books on cats and dogs is what qualified me as an expert to write for such a prestigious encyclopedia. Ironically, the censorship exercised over my entry was widely publicized in newspapers and radio stations across the nation and hundreds more letters poured in from those who supported my position. One of these from the actress Kim Basinger read as follows:

> *Dear Dr. Fox,*
>
> *I am writing to thank you for your dedication to the truth where animals are concerned. I recently read about the attention drawn by your writing for the* Encyclopedia Britannica *in the January 23, 1992 issue of the* Los Angeles Times. *As I am sure you know, the real truth scares people, especially when these same people make money off the lies of animal research. I wanted to let you, the* Los Angeles Times, *and the* Encyclopedia Britannica *know how much I appreciate everyone's attempt to speak for those that cannot speak for themselves.*
>
> *Sincerely,*
> *Kim Basinger*

This controversy illustrates how fact and truth have become disjointed, the same facts leading to one conclusive truth—that dogs have suffered in the name of biomedical progress and that the scientific validity and medical relevance of experimenting on animals today *is* being questioned—offends the truth that others live by. That truth is that the exploitation and suffering of animals is justified in the name of scientific knowledge and medical progress. The facts—to the proponents of biomedical research—of documented animal suffering are not really facts but rather unavoidable means to the vaunted ends of medical progress.

Unlike many of the people who wrote letters in support of keeping my statement intact in the 1993 revised edition of *Encyclopedia Britannica*, I actually accept using dogs and other creatures in biomedical research, but with the proviso that we learn only from those that are already sick or injured, in order to benefit them and their own kind—and our own in the process. I do believe that it is unethical to deliberately harm any creature in the name of scientific knowledge or purely human-medical progress without any reciprocal benefit to fellow creatures. No good ends can come from evil means.

The problem with biomedical research on animals, as in our other dealings with the animal and plant kingdoms, is that the boundless ethic of compassion has been replaced by the boundless imperatives of human needs and such human-centered values as progress and superiority over all life.

If this were not true, then the biomedical research community would not have reacted so vehemently to my "unbalanced" encyclopedia statement.

There is a world of difference between violating the ethic of compassionate respect for all life when we kill fleas for dogs' sake or mosquitoes (infested with malaria parasites) for people's sake, as opposed to when we turn animals into alcoholics or cocaine addicts or give restrained dogs inescapable shocks day after day as a way to study aspects of human depression. Testing a birth control or a new rabies vaccine on laboratory cats (and killing and dissecting them) and devising new hip joints for crippled dogs are arguably ethically acceptable "uses" of animals in biomedical research, because other animals and people too may benefit from the new knowledge that may come from such research.

Many people, scientists and nonscientists alike, believe that only humans and not other animals have souls, minds, and feelings. One of my early mentors, the late Nobel Laureate Konrad Lorenz, summed up the attitude that arises when there is a lack of unified sensibility as follows:

> *The fact that our fellow humans are similar, and feel similarly, to us, is evident in exactly the same sense as mathematical axioms are evident. We are __not__ able __not__ to believe in them. Karl Buhler, who to my knowledge was the first to call attention to these facts, spoke of "you-evidence."*
>
> *We have the same axiomatic certainty for animals' souls, as we have for supposing in our fellow humans the existence of a soul*

*(which means the ability to experience subjectively). A human who truly knows a higher mammal, perhaps a dog or a monkey, and will not be satisfied that these beings experience similarly to himself, is psychologically abnormal and belongs in a psychiatric clinic, as an impaired capacity for "you-evidence" makes him a public enemy.*[2]

## Agribusiness and Its Treatment of Animals

I have also experienced the lies, denial, ridicule, and concerted opposition from the agribusiness establishment that wants the status quo of intensive industrialized animal agriculture—so-called factory farming—to be maintained to further the economic interests of a powerful few. The many hidden costs are neither revealed to the public nor to shareholders. These include farm animal sickness and suffering; ecological damage; loss of non-renewable natural resources from topsoil to fossil fuels; loss of family farms and the socio-economic demise of rural communities and once-sustainable agricultural practices. The role of animal-based agriculture in contributing to increased consumer diet- and food-safety-related health problems is also denied.

I have visited factory farms and feedlots and documented how livestock and poultry are raised for human profit and consumption. Having seen alternative, more natural or ecological, farming systems in the United States and in many other countries, I do not accept food (and fur) animal factories as normal and progressive "state of the art." I regard them as pathogenic environments for the animals, and *as being symptomatic of a seriously defective human-animal bond.* To develop and accept livestock and poultry production systems that entail treating animals as mere production units and deny them their basic rights and behavioral needs is surely indicative of an aberrant mutation in human consciousness, an attitude or state of mind that lacks empathetic sensibility. It is ethically blind because it has no vestige of what I term *feral vision*—of perceiving the thing in itself, a purity and clarity of perception devoid of self-interest, that appreciates the intrinsic value of all living beings and our kinship with them.

While this mutation is such that we are neither in our right minds nor in the right relationship with animals and nature, it does not mean that we cannot

---

2. "Tiere und Gefuhlmenschen," *Der Spiegel* 47 (1980).

change our minds and behavior and recover our feral vision to establish a more equitable relationship with animals and nature: one that is based more on compassion, symbiosis, and service than on domination and exploitation. Then the possibility of building a global, humane, and sustainable society may be realized—but not until the liberation of animals from human bondage is seen as a spiritual and ethical imperative linked with our own survival and well-being.

The human-animal bond is as ancient as our earliest ancestors emerging from the dawn of time. Animals were the reference that helped us define what it means to be human and what it means to be other.

As Professor Calvin Schwabe has shown in his seminal book *Cattle Priests and Progress in Medicine*, the first veterinarians were also priests and healers of people.[3] This noble tradition of the veterinarian as a latter-day shaman, or interlocutor, between humankind and the animal kingdom is best expressed when we begin to examine the human-animal bond and seek to correct it wherever and whenever the basic bioethical principle of *ahimsa*, of avoiding harm or injury to an animal, is violated.

Animals have served us in myriad ways over millennia, and we owe them a great debt of gratitude. This gratitude we have yet to fully express in a more compassionate, egalitarian, and mutually enhancing symbiosis, since this bond today is primarily one of domination and exploitation rather than of service and communion.

Scientific research is rediscovering the many human benefits of a strong human-animal bond, or what I prefer to call the *human-nonhuman animal bond*, since we humans are animals after all. As other animals have been shown in a variety of documented studies to help people overcome great emotional difficulties and physical and psychological handicaps, so we too can help them—by saving endangered species, by restoring and protecting their habitats, and by alleviating the suffering of animals through the art and science of veterinary medicine.

## Humaneness Pays

Several studies have revealed how a positive affectionate social bond with animals helps enhance disease resistance in laboratory animals, the productivity of farm animals, and the trainability of dogs and other species.

---

3. Calvin Schwabe, *Cattle Priests and Progress in Medicine* (Minneapolis, Minn.: University of Minnesota Press, 1978).

It has now been scientifically documented that when animal caretakers express a negative attitude toward farm animals, and when the animals under their care are afraid of humans, then such animals are under a chronic state of stress. This stress has been shown to reduce the numbers of offspring that sows produce, the numbers of eggs laying hens produce, and the amount of milk dairy cows give. Meat quality and growth rates in broilers, piglets, and beef calves may also be adversely affected. In contrast, where farmers, farmhands, ranchers, and cowboys treat animals gently and with understanding so as not to trigger fear reactions and induce a chronic state of stress, and especially when there is a strong social bond between caretaker and animal, the animals are healthier and more productive and profitable. In sum, humaneness pays.

But this evident fact may be discounted in favor of the cost-savings of developing large-scale, labor-saving, intensive livestock and poultry systems where one person is in charge of hundreds, even thousands, of animals. Such labor-saving "efficiencies" have been criticized for decades by those who believe that farm animals need individual attention or at least regular daily inspection. It may be argued that a good stockperson with a positive attitude toward animals on a large factory farm will do a better job than one who is indifferent, or one the animals fear. However, working in large intensive production systems has been shown to adversely affect the behavior and attitudes of stockpersons—notably causing increased aggressive behavior, which has a detrimental effect on the animals' welfare and productivity.

There is a distinction between humanely directing an animal's *telos* (final end or ecological purpose) for one's own benefit, and disregarding both telos and *ethos* (or intrinsic nature) and manipulating both, as in factory farming and genetic engineering, for exclusively human benefit. The good farmer and pastoralist knew how best to profit from the telos of plants and animals without harming either their ethos or the *ecos*—the ecosystem—as a creative participant or symbiote. This distinction between kindly use and selfish exploitation is the difference between sustainable and nonsustainable living. The difference between treating animals and other life forms as ends in themselves, respecting their ethos and ecological role, and of using them as a means to satisfy purely human ends is a central issue of bioethics.

The consequences of this utilitarian attitude and relationship are potentially harmful economically, environmentally, socially, and spiritually. The keeping of wild animals in captivity for public entertainment, the selective breeding of genetically defective "bonsai" pets, and the genetic engineering of animals for various commercial purposes also illustrate how one-sided the human-animal bond has become.

In sum, natural living, like natural or organic farming, has its own integrity. Ironically, in the process of "denaturing" animals under the yolk of industrial-scale domestication, we do no less to ourselves and suffer the consequences under the guise of "civilized" necessity and progress. But there is nothing civilized or progressive in this utilitarian attitude toward life, since it ultimately reduces the value of human life and all life to sheer utility, which is the nihilistic telos of technocratic imperialism and materialistic determinism.

Until we fully address and repair the human-nonhuman animal bond *and this is, I believe, the primary task of veterinarians and humanitarians*, the world will continue to suffer the adverse consequences of a pathological human attitude toward fellow creatures and nature.

A society that destroys nature's biodiversity and resources will become materially bankrupt. Similarly, we will become spiritually bankrupt if there is no collective regard for the intrinsic value and ethos of animals, and of their telos or purpose in the ecological community. It is this Earth community, not simply the human community, that we must all come to serve if we are to survive, prosper, and evolve.

## Cats and Human Relations

Cats, like other animals, can evoke a variety of emotions, both good and bad, in us often-not-so-rational human beings. In my role as a veterinary consultant through my syndicated newspaper column "Animal Doctor," I learned of a woman who became increasingly unsatisfied with her relationship with her husband to the point that whenever he made any advances toward her, she would put the cat between herself and her husband and start to pet and talk about the cat instead. The husband became so furious and frustrated at the way his wife used the cat as a way of avoiding him that he wrote to me in desperation, saying that he was so jealous of the cat that he wanted to kill it. He described his wife's behavior in his letter but clearly had not made the connection that his wife did not necessarily love the cat more than he; rather, she was using the cat as a means of avoiding closeness with him. I advised them both to see a marriage counselor.

Most bizarre of all was a six-page letter from a distressed lady who was convinced that all the cats in the neighborhood were plotting against her.

"Whenever I pass a cat, it always looks at me. And when I see two or more on a neighbor's step, they always stare at me and I can tell they hate me and are plotting something. But why me? I love cats and never harmed one in my life." The letter rambled on with her describing in detail her paranoia, justified, she felt, by the fact that whenever she was thinking about cats scheming to harm her, she would soon run into one.

While flying to Amsterdam to talk to the press at the 1981 International Cat Show, I met on the plane an American woman who did not like cats. She told me she was not afraid of them, liked animals in general, and didn't mind being touched by cats. But she couldn't stand their *purring*. I asked her why, and she said it irritated her. "Was it too intrusive, like someone massaging your insides?" I pried. "Yes!" she exclaimed. "It's funny you should put it that way." Subsequent conversation on the long flight confirmed my suspicion that she was afraid of real closeness. She had been abused as a child, and her third marriage was on the rocks.

In my work in animal protection I am constantly appalled by cases of cruelty by children toward animals: children throwing kittens against a wall, tying a tin can onto a cat's neck or tail, or soaking a cat's tail in gasoline and setting fire to it. Such a pathological lack of empathy, often combined with aggression redirected onto a harmless creature, is a serious sign of emotional disease. Studies have shown that many adult sociopaths have a past history of cruelty toward animals and almost invariably were victims of parental abuse during their formative years.

A prison psychologist recently told me about one tragically imbalanced man serving a life sentence in an Alabama jail. The man refused to wash himself, was filthy, depressed, homicidal when handled, and was hosed down every other day to be kept clean. The psychologist, familiar with pet-facilitated therapy, gave the inmate a little kitten to care for. Initially the man tried to ignore the kitten, but she quickly won him over, opening the man up to life itself. By the next day the man was keeping himself clean and tending to the kitten; soon thereafter, he came out of his depression and began to relate to the wardens. His homicidal fury slowly burned out.

Another cat therapist was a lifeline out of insanity. This same psychologist recounted that only a few days after he gave a young cat to a schizophrenic patient who had spoken to no one and avoided all human contact for fourteen years, she began to talk to attendants and fellow patients.

Mental health specialists, sociologists, veterinarians, and others are now examining the many benefits of this human-animal bond. I feel a sense of vindication as society comes to accept what the animals taught me since childhood: that they can be our healers, as well as our best friends and teachers.

I remember almost weeping when I saw a news clipping of an old lady with her cat. She was being evicted and put in a retirement home and did not want to be separated from her cat. She said that her cat was "the only living thing that she had touched over the past twelve years." I knew that this old woman would probably die soon after she was separated from her cat, or would withdraw into depression and lose all interest in life. I have urged retirement homes and hospitals for the chronically ill and severely handicapped to recognize that a pet on the ward will not bring in dirt and disease but a light into the lives of all the residents—and that every effort should be made to keep people and their pets together.

This reminds me of a letter that was sent to President Franklin Pierce from Chief Seattle of the Duwamish Tribe, which included the following words of wisdom:

> *What is man without the beasts? If all the beasts were gone, man would die from a great loneliness of spirit. For whatever happens to the beasts, soon happens to man. All things are connected.*

It was difficult for my own children to cope with the loss of their cat Lily, who was getting very old and simply disappeared one day. Five-year-old Camilla hoped fervently that Lily would return someday and held onto that hope for many months. It would have been much easier for her had she seen Lily's dead body. Then she would not have used the hope of her cat returning as a way of coping with her sense of loss. Had she seen Lily's body, such wishful, magical thinking would have been dispelled. I wish that parents would not try to protect their children from such loss by not allowing the child to see the dead pet. I advise all parents to help the child cope with such loss and separation not by telling the child that it was only a cat, or that they will get a new kitten as a replacement soon. Statements like these can make the child hate the parents and reject the new pet. Loving parents can do more harm than good when trying to make the child's hurt go away. Simply respecting the child's deep loss and allowing the child to mourn with kind and supportive words of understanding is all that is needed.

My son, Mike, Jr., could be hard to reach, since like me when I was younger, he used his intellect to stop himself from feeling the hurt of loss.

He loved Lily dearly, and while he knew she was dead and would never return, he repressed his tears. Then some time after Lily had disappeared, he had an argument with his sister and felt rejected when I intervened on her behalf. He went to his room and began to cry. He tried to hide his tears when I came in to see him, but the flood-gates burst as he wept for Lily. His feeling of being rejected by me and his fear of loneliness had brought up the emotions that he had repressed for so long over Lily's death. We wept together and talked about how wonderful Lily was, what a good life she had, and how it hurts to lose what you love. Young Mike soon recovered and remarked that since he'd been told that people go to God's heaven and animals to Jesus's heaven, he will go to Jesus's heaven when he dies so he could be with all the animals, and with Lily. This was an interesting rationalization by a seven-year-old and quite spontaneous too, since we had never talked about where animals go when they die.

It intrigued me that he should think that animals go to a different heaven, and I blame the insidious cultural influence on all our thinking, of the mainstream Christian belief that only humans have immortal souls and go to heaven. As the Book of Ecclesiastes states, "Man has no pre-eminence above a beast: for all is vanity. All go unto one place: all are dust, and all turn to dust again. Who knoweth that the spirit of man that goes upward and the spirit of the beast that goeth downward to the earth."

In a Buddist monastery, high in the mountains of Tibet, a translation of an early Christian document was found. It had been taken there by people of the so-called Essene community who hid it from the hands of corrupters.

The document is called "The Gospel of the Holy Twelve." In one of the chapters of this particular work is the following legend. Whether one believes in its authenticity or not, this story, in its very simplicity, says all that needs to be said about our attitude toward those of different species.

*As Jesus entered a certain village He saw a young cat which had none to care for her, and she was hungry and cried unto Him, and He took her up, and put her inside His garment, and she lay in His bosom. And when He came into the village, He set food and drink before the cat, and she ate and drank, and showed her thanks unto Him. And He gave her unto one of His disciples, who was a widow whose name was Lorenza, and she took care of her. And some of the people said, This man careth for all creatures . . . Are they His brothers and sisters that He should love them? And He said unto them, Verily these are your fellow creatures, of the great Household of God; yea, they are your brothers and sisters, having the same breath of life in the Eternal. And whoso careth for one of the least of these, and giveth it to eat and drink in its need, the same doeth it unto Me; and whoso willingly suffereth one of these to be in want, and defendeth it not when evilly treated, suffereth the evil as done unto Me.*

There are great cultural differences in people's attitudes toward suffering and the value of life itself. As a veterinary student spending one summer in Morocco, I was shocked to find an abandoned, starving little kitten mewing for food and affection beside a Moslem shrine on a busy side street in Rabat, being ignored by all who passed by. Such indifference toward the suffering of others contrasts the poor old black woman I saw in St. Louis, who spent a small fortune feeding the homeless cats in her neighborhood. No doubt there were a few good souls in Morocco who did no less for such helpless creatures. I wonder, is such benevolence and compassion devoid of self-interest? People need to feel needed; also, they identify their own sense of helplessness and vulnerability with the plight of animals. But regardless of selfish motivation and identification, such active compassion is superior to passive indifference. It stands out in sharp contrast to a psychologist I know who used electrical shocks to force cats to drink alcohol and become addicts, a purported animal "model" of alcohol addiction in humans (research paid for by taxpayers). While the brain and biochemical changes associated with alcohol addiction might be similar, the *causes*, the reasons why people become addicted to alcohol, are quite different. Therein surely lies the cure, making such brutalization of cats, and of the sensibilities of the scientist, pointless sacrifices.

This reminds me of the pagan European practice of burying a cat alive to ensure a good harvest and of sealing a live cat into one wall of the cottage to keep away evil spirits. Such superstitious nonsense this is, motivated by fear of the unknown and a lack of rational understanding.

The persecution of cats, along with people who loved their cats, is historically well documented. For instance, in 1484, Pope Innocent VIII (1432–1492) formally sanctioned on behalf of the church that "sorcerers" and "witches" had the power of evil. He also ordered that pet cats be included when witches were burned at the stake. As late as the seventeenth century in many parts of Europe, it was risky for an old woman to have a cat as a pet. My friend Lewis Regenstein in his book *Replenish the Earth* writes that:

*The belief that Satan could and did assume the form of animals was widespread during the Middle Ages. Black cats were the subject of particular suspicion, frequently becoming a symbol of witchcraft. Many cats were summarily put to death and people often refused to own them.*

*The ceremonial burning of cats was a tradition on certain religious holidays, especially in France. These customs are described in detail by the eminent Scottish anthropologist Sir James George Frazer (1854–1941) in his classic work on primitive cults, myths, magic, and religion,* The Golden Bough. *In Alsace, cats were burned in the Easter bonfire; in the Ardennes, on the first Sunday in Lent; and in the Vosges, on Shrove Tuesday. In the festive midsummer fires, cats were roasted over the bonfire at Gap, in the High Alps; and at Metz, a dozen cats were placed in wicker cages, and "were burned alive in them, to the amusement of the people." The Europeans paid dearly for their persecution of cats. The elimination of these predators of rodents brought about a proliferation of flea-carrying rats, which helped cause and spread the bubonic plague, or Black Death, that decimated Europe in the late 1300s, killing off nearly half of the population between 1347 and the end of that century.*[4]

Clearly we have come a long way in extending compassion and rights for animals, but there is scope for further moral progress in every nation. In the final analysis, this is enlightened self-interest, because compassion and reverential respect for all life are surely the hallmarks of greatness for every nation.

---

4. Lewis Regenstein, *Replenish the Earth: A History of Organized Religion's Treatment of Animals and Nature* (New York, N.Y.: Crossroad Publishing Company, 1991).

# 13

## | My Rabbit and the Light: Some Personal Reflections |

Companion animals can have a profound influence on young children, even when they pass on. I had a pet rabbit when I was about three years old. Thumper was a dwarf Dutch bunny—gentle, warm, and quick. He was housed in a fine wooden hutch my father had crafted, with a warm nest box and a wire-enclosed porch and walkway.

I would go and see him every morning before breakfast and always had something to tell him or a question to ask him. One morning I went to greet him and found him lying flat on one side, as though he were asleep. I had never seen him behave like that before. I called his name, but he did not move an ear or open his eyes. I was scared. I opened his hutch, and found a stick to gently touch him with. He didn't move. I felt him with a shaking hand. His body was cold and no longer moved when I pressed his thigh gently.

I was bewildered, rather than sad and tearful. Those feelings came later. The evening before he was so alive and well, but now he was gone. But what was gone, I wondered, since his body was still there, as though simply asleep. Maybe he could be warmed in the sun.

I ran to tell my mother that Thumper was dead and she must come and help. His death was confirmed, and we buried him after a little ceremony in the garden later that morning. From then on, cold and death were linked in my mind.

It was a mystery to me, the question of where the life of Thumper, which once animated his body, had gone, leaving his body behind. I had heard about Heaven and someone told me that Thumper would be there waiting for me when I died. But I wondered how could he be? In what form would he appear in, since his body was buried in the garden? I knew that the skin and flesh would be taken into the soil by a host of insects, leaving only his bones, for I had often found the bones of little creatures while playing in the grass and bushes, and maggot-filled, cat-killed voles and songbird carcasses. His body returned to the earth, so where did the rest of him go?

A few days later, with Thumper still alive in my mind, I went into the back garden early one summer morning. The flowers were head-high and the mist was thick with their perfume and the buzz of myriad insects. As I breathed in this fragrant light and looked up through the mist into the heart of the hazy sun, I had a frightening but exhilarating experience and one that remained with me forever. Suddenly I became a mere particle in that misty light. I felt connected with every particle that light embraced, and its embrace was infinite. I didn't see Thumper in the light. I didn't need to. The warmth of the morning sun reminded me of the warmth that my parents had always given me. But it was so much more intense. I felt it enter into the very core of my being.

It was all knowing and all loving and made me feel infinitely secure in its embrace. I no longer missed my friend Thumper because he was part of the light. I saw and felt how every living being in the garden, every flower, herb, shrub, insect, and bird was part of this light. It was within them and around them all. In that moment of realization, I had no sense of my own self being separate from anything else. From that time on, I never feared death or dying. Even during some of the most difficult, heartbreaking events in my life, I would find strength and trust in life by thinking about Thumper and how his light and warmth had returned to the source of all being from which all things arise.

Many years later I thought of Thumper as I mercifully stomped dozens of diseased rabbits to death. It was one of those perfect Saturdays in early June, ideal for a walk over the moors and through the limestone dales of Derbyshire. I let the wind take me wherever it would, feeling it pushing me up the hills as I kept my back to it all the time. I felt as though I was flying, my body, thanks to the wind, keeping up with my soaring spirits. I was elated because I had the letter in my pocket that I had been waiting for. It notified me that I had been accepted for the 1956 first academic year at the Royal Veterinary College, London. My childhood dream was coming true. I had always wanted to be an animal doctor, and if I passed all the examinations, I would be qualified in five years' time.

Five years, I was thinking. For the past five years, I had already been helping local veterinarians on their farm and on house calls. How much more to learn—to really know how to treat sick animals, diagnose their ailments, perform surgery, deliver calves. Five years seemed like a long time, and even longer if I failed some of the exams. I groaned to myself as I scaled an old limestone wall, startling a ewe and her young lamb who were napping in the sunlight on the other side. I apologized to them and they trotted off, bleating briefly before stopping and turning to give me a look of curious disdain. The wind pushed me up the hill and at the top I breathed in deeply, enjoying the day and the boundless promise of exhilarated, youthful spirits. I readied to race down into the dale below, noting the sheep trail between the limestone rocks and boggy reed beds that I needed to skirt to avoid a bad tumble. Then I saw a rabbit moving beside an outcrop of limestone. She was not moving normally and appeared drunk and dazed. Perhaps she had been poisoned, I thought, as I moved slowly and cautiously toward her.

She disappeared behind the rocks and I followed her into an open glade that was pocked with rabbit holes. It was a large warren, but the eerie thing was that she just kept wandering around and didn't bolt down one of the holes. She seemed oblivious to my presence.

Then I saw rabbits wandering aimlessly and dying slowly by the score in the bright sun of this warm summer's day. The innocence and verdant beauty of this limestone dale was shattered. There was no breeze, and the buzz of carrion flies filled the heavy air. I walked slowly and quietly, so as not to frighten any of the less sick rabbits who seemed vaguely aware of my presence, but made no effort to escape. One so blindly sick, hobbled toward me, without seeming to be aware of my presence. Or was she asking me to kill her?

I picked her up. Bloody mucous from her mouth and gasping nostrils splattered my face and parka jacket as the pathetic creature struggled weakly in my hands. Her eyes were so swollen I felt they might burst. I could not save her, nor any of her other dozens of warren relatives, who were circling, crouching, and convulsing, as far as my eyes could see. I quickly dispatched the poor creature with a "rabbit punch" at the base of her skull. I went on killing automatically until my hand hurt too much from striking and dislocating their necks. So I used the sharp heel of my boot, placing each rabbit on a rock so that the skull would implode instantly rather than sink, possibly uncrushed into the soft ground.

Eventually there were no more rabbits alive and I was surrounded by the carnage of my own compassion. Perhaps the rabbits stayed aboveground to save others in the warren from becoming infected. I doubted, however, that there were any survivors. I sat down in their midst as my senses slowly returned. I heard a skylark's spiral song rising above some distant meadow as the breeze returned and rustled in the reeds close by.

Needing to vent my rage at the plight of the rabbits, I walked and walked until I could walk no more and collapsed on the wide upland moor, lying with my back on the soft, warm earth. I let my mind lift into the clear sky, following the cadence of motion and sound so perfectly balanced in the spiraling and trilling of the larks. I was able to weep then for the tragedy behind and below me: so many slow-dying rabbits with heads swollen and protruding eyes from a disease that I had read about—myxoematosis, or rabbit plague. This disease had been deliberately introduced by government agents to reduce the rabbit population. It ensured long suffering before death, leaving behind a resistant population that gradually increased in numbers. But unaccounted numbers of foxes, owls, and hawks had died from starvation when the rabbit warrens were decimated by this disease. A few years after this mass extermination of rabbits, they became even more of a problem because these natural predators had died of starvation and were not around to help keep their numbers in check.

Such a cruel, futile, and perverse attempt at biological warfare by man had made the valley, and dozens more, a festering trench of blind, starving, disoriented, and dying rabbits. What right had I to end their suffering? Better perhaps to let the disease take its course and leave the valley. But their suffering was mine, and I would have been a coward to have left them alive.

On my way home, the evening breeze brought its cool sweetness from across the distant moors and my sadness turned again to the rage and humiliation (for being human) that I had felt as my hand and boot snapped and crushed life out of rabbit after rabbit. I remembered then that none of the rabbits had screamed: their death was perhaps a merciful release from their silent suffering. The scream of a rabbit in pain or fear I know was a penetrating and distressing emotional tone, at least to my ears. I caused them neither pain nor fear when I assumed responsibility for their suffering that day. Perhaps it was my first real examination, to test if I really had the right qualities and potential to become a veterinarian and, first and foremost, to be able to put compassion into action.

# Part Two
## Cat Body

# 14
## | Trends in Companion Animal Care, Health, and Welfare |

## Animal Care and Welfare: A Snapshot

A recent American Pet Products Manufacturers' Association National Pet Owners' Survey conducted for the $31 billion pet product industry includes some basic, as well as some interesting and unsettling, information.

*Demographics*
A total of 64.2 million U.S. households own a companion animal, up over 10% from a decade ago. There are 77.7 million cats. More cat owners, six in ten, keep their cats indoors; one in ten cats are kept outdoors; the remainder, three in ten, are indoor-outdoor cats. There are 65 million companion dogs. Two out of ten dogs are kept outdoors.

*Spay and Neuter Statistics*
More than eight in ten cats are now spayed or neutered. Nearly thirty states have laws mandating that animals adopted from shelters be spayed or neutered.

*Nutrition Statistics*
Dry food is the most widely used cat food, leading to more serious health consequences, including obesity, diabetes mellitus, cystitis, and the feline

urologic syndrome (when high in starch). Semi-moist packet cat foods, which are high in sugar as a preservative, can also be problematic.

Fewer than 10% of cats are obese or overweight, a decrease from 16% found in the 2000 survey. More dogs (16%) are obese or overweight, up from 12% found in the 2000 survey.

Cats and dogs are receiving better nutrition, according to the 2003–2004 study. More people are feeding their pets higher-quality commercial pet food, as well as home-prepared diets, which I regard as a significant improvement.

## Animal Protection

Over thirty states now have laws making animal abuse a felony. Six states have laws granting immunity from civil or criminal liability to veterinarians reporting suspected animal cruelty.

More judges, prosecutors, and social workers are aware of the link between family, animal, and spousal abuse—as well as the link between childhood cruelty toward animals and violent criminal behavior later in life.

As of 2000, only three states mandated pound seizure (allowing cats and dogs to be taken from shelters by research labs for experimentation). More than a dozen states now prohibit this practice in most municipalities and dropped the practice because of public outcry.

Both common and statutory laws are now increasingly acknowledging the loss of an animal companion as a basis for recovery of damages above and beyond the fair market value of the animal.

## Euthanization Statistics

According to the Humane Society of the United States (HSUS), animal shelters are euthanizing fewer animals. An estimated 4.6 million (4.5%) of the 120 million dogs and cats owned are euthanized at this time (2005–6), down from 5.6 million (5.5%) of the 110 million dogs and cats owned in 1992.

## Modern Medicine Issues

The HSUS *State of the Animals 2001* report stated that kidney transplants for cats were noted as an advance in veterinary care. I am deeply concerned about the lack of consideration for the source and fate of donor cats. This topic is hotly debated in the veterinary journals in England. The consensus in England is that it is not acceptable to perform kidney transplants on cats.

The view in the United States is that if shelter cats are going to be euthanized if they are not adopted, then they should be "used" as transplant donors. I believe that this logic is ethically questionable. Furthermore, veterinarians have reported that cats with kidney transplants done because the cats were in kidney failure are likely to develop secondary diabetes mellitus.

*Roam Free?*

The insistence of many cat owners and some animal rights advocates that domesticated cats be allowed to roam free needs to be confronted. Some European countries accept letting cats run free, contending that it is their right to live and roam naturally.

This practice has a serious adverse environmental impact. Allowing domesticated cats to live outdoors leads to the killing of wildlife and puts the cats' own lives at risk from infectious diseases; some of these diseases can be spread to wild mammals, including endangered species like the Florida panther.

The outlawing of free roaming domesticated cats is a very different issue from finding humane solutions for feral cat problems. Programs advocating spaying and neutering, vaccinating, and releasing of feral cats with daily provisions of food may be acceptable, only if there are no wildlife at risk from feral cat predation and disease.

# Animal-Based Industries

The high regard in the West for certain animal species, which are variously regarded as food or vermin in other cultures (most notably, cats and dogs), is an irony that some critics see as sheer overindulgence and misplaced emotion—particularly when billions of dollars are spent annually on the care of such animals. That such animals are considered human companions and family members has helped promote the social acceptance in the West of animals having rights and being worthy of moral consideration. This is seen as a significant threat to various animal-based industries in the United States—particularly the factory-farming, fishing, sport and trophy hunting, trapping, animal research, and wildlife trade and fur industries.

These animal industries erroneously dismiss pet lovers as sentimentalists who anthropomorphize their animals and have no real understanding of what really goes on in these "vital" industries.

These exploiters of animals do not wish to accept the fact that animals have feelings and scientifically documented, emotion-mediating neurochemical pathways virtually identical to ours (as documented in Chapter 10), because to do so they would have to face their own moral dilemma of profiting from cruel forms of animal use and abuse.

We are now seeing an about-turn, a radical change in human-animal and Earth relations, with the realization that we are part of the Earth and the Earth is part of us. When mankind harms the Earth, mankind harms himself. When mankind demeans animals, human rights are ultimately eroded as well.

## Commercialization Issues

While the conscientious veterinarian tries to balance the interests and needs of the client and the animal (often a difficult challenge), market research by the pet food, pharmaceutical, and supplies industries are primarily focused on selling ever more products and services.

From drugs, insecticides, and snacks to toys, training collars, and invisible fences, critics see these companies catering more to owner/caretaker/guardian needs, fears, and phobias rather than to animal needs and best interests.

All these insecticides that are injected, fed to, and put on companion animals finish up in the environment via their stools and urine. This effect profoundly affects the entire ecosystem. It is imperative that the government address this serious issue.

## Vaccination Practices

Until recently, dogs and cats have been given far too many vaccinations—and too often. While many veterinarians are now changing their vaccination protocols, and owners are becoming more aware of the risks of vaccination, too many kennels and catteries that board animals still insist on up-to-date vaccinations. A safe alternative to unnecessary repeat-vaccinations is to have blood titers taken to determine if any revaccinations are needed.

Annual "booster shots" are now being shown to often cause more harm than good. They have often resulted in a variety of autoimmune diseases, chronic immunodeficiency and endocrine disorders, and, notably, cancer (fibrosarcoma) in cats. (See pages 164–66 for more details on this important topic.)

*The Dark Side*

Many people enjoy a deep sense of communion with their animal companions. This special relationship has led to a plethora of books celebrating this bond, as well as the spirituality of the animal connection and its healing and transformative powers. Companion animals are even regarded as angelic beings.

But there is a dark side to our relationship with companion animals that love and respect alone will never eliminate.

Education, legislation, investigation, and prosecution are called for when addressing the cruelty suffered by animals in a society where closed-door institutions like the pet food industry contract out to test animals to develop, for example, a new diet for cats with kidney disease. This research is all too often conducted at the animals' expense.

The public is not aware of how such research is carried out. Most of the cats' kidneys are surgically removed, and experimental dietary products are tested. The "special" diets that are developed are done at many animals' expense and suffering.

I do not believe this is the way to proceed to sell market commodities. The pet food industry obviously disagrees.

That a scientist, lab technician, or student can design and perform experiments on cats and dogs that outside of the institutional setting would be a violation of anti-cruelty laws points to the power of situational ethics over the ideal boundless ethic of compassion.

Laboratory animals are being bred with specific diseases to facilitate development of new drug tests and screens for genetic disorders. Modern science is experiencing an explosion of biomedical research, especially on genetically engineered rodents.

Genetic engineers are now eyeing the purebred dog as the next laboratory rat. The U.S. government has provided researchers with $50 million of taxpayer dollars to sequence the canine genome for the purpose of identifying genetic abnormalities (of which there are close to four hundred) that are analogous to those seen in *Homo sapiens*.

Dogs will be deliberately bred to have genetic disorders; while this may help breeders eliminate defects from the gene pool of afflicted breeds, it is clear that dogs will be made to suffer in this new service for humanity as models of similar hereditable diseases in man.

I have advocated for years that sound science can demonstrate that animal experimentation is not necessary.

It is ethically wrong, and, I believe, unnecessary to deliberately induce illness and injury in healthy animals. There already exist many sick and injured animals. More ethical animal research can be done in collaboration with veterinarians and their clients.

With good liaison and collaboration between veterinary schools and veterinarians in private practice, much can be learned from testing such new diets, surgical procedures, medications, and diagnostic procedures (with the owners' informed consent) on animals who are in need of such advanced treatment.

I firmly believe that these methods are the way to progress humanely toward better companion animal health. Much unnecessary laboratory animal research would be eliminated.

## The Essence of Companion Animals

One of my classmates, the late Derek Pout, a British veterinarian, wrote to me, "Animal lovers seem to love animals for their own personal needs, rather than for the animal and all its qualities." This is, I believe, somewhat an overgeneralization. I am finding that more people are much more responsible animal care-givers. They extend themselves to provide their animal companions with not only the best nutrition and veterinary care, but also an environment in which they can enjoy being animals. They thus allow their companions to experience their own dogness, catness, guinea pigness, birdness, and fishness; animals are no longer simply regarded as emotional surrogates and extensions of their human care-givers.

Animals are not just intelligent beings; they also have feelings. I was recently asked in a television interview, "What is the most highly developed ability of companion animals, Dr. Fox?" I answered, "Their ability to empathize."

We must be very mindful of the emotional "vibes" we emit to these highly evolved, very empathic beings. Just like humans, they have emotional intelligence.

Categorically, animal rightists believe that animals should not become a means to human ends because they have their *own* lives, interests, and ends in and for themselves. For this reason, it is argued that keeping animals as pets is morally wrong. This is a contention that must be addressed.

Humans enjoy mutually enhancing, symbiotic relationships with other animals. We should really celebrate this bond; but we must develop and nurture it with understanding and education. A healthy human-nonhuman animal bond is, by its nature, mutually enhancing; each being meets, in part, the other's interests and ultimate well-being.

### In Our Companion's Best Interests

Fortunately, many of my close colleagues are holistic veterinarians. They are a new breed of professional animal health and welfare advocates, focusing not just on physical manifestations of animal health, but the emotional and social aspects as well.

These colleagues stress the following principles for basic animal companion health.

1.  Get a mixed-breed dog or cat; adopt a healthy companion from the shelter. The in-breds and purebreds generally come with a host of genetic defects unless you can find a reputable breeder who keeps impeccable progeny testing records.

2.  Provide a healthy social and emotional environment, including compassion and respect for a species with needs different from our own.

3.  Provide healthy (organic, if possible) nutrition. Avoid the highly processed foods containing additives and condemned animal parts.

4.  Obtain holistic veterinary treatment and set up a health-care and maintenance program that means a regular (ideally every six months) consultation with a veterinarian who will spend time talking about optimal care, nutrition, minimal vaccination, and ways to deal with and prevent common health and behavioral problems.

### Thoughts on Progress

The movement to protect companion animals has made considerable progress since I first became involved more than thirty years ago. Much of the credit goes to the public: its concern for the welfare of animals, both through individual efforts and through the support of the local shelters and the various humane societies, has blossomed. Improved shelter facilities, well-trained and more dedicated staffing, behavioral counseling programs, spay/neuter and adoption programs, improved anti-cruelty investigations, and stricter law enforcement, coupled with more humane education outreach programs in schools throughout the country, are to be applauded.

Society is serving the needs, rights, and interests of the animals with which it coexists. Their time has come.

## Onward

The social environment and relationships of dogs, cats, and other companion animals is of continuing concern to me.

Owners must appreciate the fact that having two animals, be they cats, guinea pigs, parakeets, or goldfish, is more humane than simply having one animal that is left alone during long work days, living a life of extreme social deprivation.

One must avoid the genetically engineered "Frankenpets"—a startling example being genetically engineered fish that glow in the dark.

Cats, dogs, sheep, and other animals are now being cloned. The unknown health ramifications of these experiments with nature are frightening. There is a repugnance factor here; the general public has a "gut feeling" that this aspect of commercialized bioscience is morally and ethically wrong.

Before we can hope to move federal and state governments to enact more effective animal and environmental protection legislation and enforcement, we must educate the legislators to see the connections, the big picture. They must acknowledge that the plights of the natural environment and exploited animals mirror the plight of the human condition, with its lack of compassion and respect for life, both human and nonhuman.

The greatness of a nation can indeed be measured by the status it has accorded to every life, human, nonhuman, plant, and animal.

Just as the security of a nation is tied to the fate of the Earth, so the future of civilization is tied to effective education, legislation, and law enforcement that functions to protect all animals, the natural environment, and the last of the wild.

But as long as other animals—from puppies and kittens, to parrots and potbellied pigs—are regarded primarily as commodities, and there are no laws to prohibit lawmakers from winning elections funded by those vested interests responsible in part for the holocaust of the animal kingdom, the ethical and spiritual erosion of society will continue unabated.

Evil flourishes where good men do nothing.

# 15
## | A Holistic Approach to Health |

The proper care of a companion animal, the provision for its basic social and emotional needs, coupled with appropriate training will do much to "nip" behavioral problems in the bud. The result would be far fewer animals being abandoned, put up for adoption, or euthanized.

Animal shelters provide extension programs and hotlines to promote proper care and understanding, as well as counseling when behavioral problems arise. These valuable programs are increasingly recognized preventives to newly adopted animals being returned by owners.

From March 2004 to March 2005 I kept a record of the hundreds of different health and behavior problems that readers of my nationally syndicated "Animal Doctor" newspaper column wrote to me about. Many of these problems had received veterinary attention, and for various reasons had not been resolved.

The top five health problems for cats were:

1. Itching, hair pulling, raw spots, and hair loss

2. Feline urologic syndrome (urinary tract inflammation and calculi/ blockage of urine)

3. Chronic diarrhea, possibly inflammatory bowel disease

4. Kidney failure, adverse reactions to vaccinations and anti-flea medications

5. Repeated vomiting of food

The list of the five most common behavioral problems in cats were:

1. House soiling, most often with urine (and not associated with FUS), but frequently with feces as well

2. Aggression toward humans (in some cases possibly linked to hyper-thyroidism)

3. Aggression toward other cats in the home

4. Biting too hard during play and "love biting" while being petted

5. Addiction to dry food, senile dementia, and spraying by mainly neutered males

Most of these health and behavioral problems, as I have long advocated, can be prevented. In many instances, this can be achieved by proper diets (see later) and by a more holistic approach by veterinarians who include behavioral counseling and preventive health-care education for cat and dog owners—especially those with a new kitten or puppy. Many animal shelters are now doing this, and veterinary colleges are including this in their curricula and in students' in-field experience and community outreach.

There are genetic susceptibilities of certain breeds of dogs and cats, individual genetic abnormalities notwithstanding. However, most of the above health and behavioral problems—which result in much animal suffering and emotional and financial cost to the primary care-givers/owners—should and *can* become something of the past when the basic principles of holistic veterinary preventive medicine and responsible animal care are applied to the companion animal–human relationship. These translate into care-givers/guardians/owners providing animals with not only the right nutrition, but most importantly with a right relationship based on an understanding of the animal's behavior and emotional, social, environmental, and physical needs, which boils down to providing the right environment for the animal.

The other right of all animals under our care is proper veterinary care, which must begin with a more conservative approach to prescribing and administering potentially harmful and often unnecessary vaccinations and other drugs, especially those used to control fleas and ticks. These are wrongly toted under the banner of "preventive" medicine for pecuniary if not misguided reasons.[1] It is not unlike the wholesale use of antibiotics in intensively farmed animals that puts consumers at risk worldwide from antibiotic-resistant strains of bacteria in contaminated meat, eggs, poultry, and dairy products.[2] (Those who have read Ivan Illich's book *Medical Nemesis*[3] may well see a parallel in this veterinary nemesis that I have been seeking to rectify for many years. And I am not alone in this deep concern for the animal health and well-being, for the good of every society, East and West.)

## Cat Vaccination Protocols

The practice of giving dogs and cats several different vaccinations against various diseases all at the same time early in life and then again every year as "boosters" for the rest of their lives is coming to a close. This is for two primary reasons: animals can have adverse reactions to vaccinations that can impair their health for the rest of their lives; routine booster shots are not

---

1. Adverse vaccination reactions resulting in disease (so-called vaccinosis) that have been identified or are highly suspect in dogs include encephalitis, seizures, polyneuropathy (weakness, incoordination, and muscle atrophy), hypertrophic osteodystrophy (shifting lameness and painful joints), autoimmune thyroiditis and hypothyroidism, liver, kidney, and bone marrow failure variously associated with autoimmune hemolytic anemia, immune mediated thrombocytopenia. Certain breeds are more susceptible than others. Documentation on feline vaccinosis is less complete, but suppression of the immune system and increased incidence of chronic infections and allergies are probable consequences of some feline vaccinations, and of cats being given too frequent and unneccessary vaccinations. The increased recognition of vaccinosis means a more conservative approach to vaccinating dogs, cats, ferrets, and other companion animals.

2. For more information contact the Association of Veterinarians for Animal Rights (PO Box 208, Davis, CA 95617-0208) and the Holistic Veterinary Medical Association (2214 Old Emmorton Rd., Bel Air, MD 21015). For beginners in the quest for a more holistic approach to companion animal care see: Donna Keller's book *The Last Chance Dog* (New York: Scribner, 2003); Richard H. Pitcairn and Susan Hubble Pitcairn's *Natural Health for Dogs and Cats* (Emmaus, Penn.: Rodale Press, 1995); Franklin D. McMillan's *Unlocking The Animal Mind* (Emmaus, Penn.: Rodale Press, 2004). More advanced texts for veterinarians include: Allen M. Schoen and Susan G. Wynn's *Complementary and Alternative Veterinary Medicine* (St. Louis, Mo.: Mosby, 1997); Susan G. Wynn and Steve Marsden's *Manual of Natural Veterinary Medicine* (St. Louis, Mo.: Mosby, 2003); Cheryl Schwartz's *Four Paws and Five Directions* (Berkeley, Calif.: Celestial Arts, 1996).

3. Ivan Illich, *Medical Nemesis: The Expropriation of Health* (New York, N.Y.: Random House, 1976).

needed since earlier vaccinations have given animals sufficient immunity to the diseases in question.

First, very young (i.e., before twelve weeks of age) kittens should not be given vaccinations since this can interfere with the natural immunity in their systems conferred by the colostrum, or first milk, of their mothers. But for kittens and young cats in an animal shelter environment, a more rigorous vaccination protocol than I outline below is called for, especially because of exposure to potentially sick and infective cats, and too often, inadequate quarantine. Adult animals in a compromised immune state should not be vaccinated, including: those who are ill, injured, or being given an anaesthetic and operated on for spaying, castrating, or for any other surgical procedure; those who are pregnant or nursing; those who are old and infirm.

The following protocols for vaccinating cats have been published in the *American Holistic Veterinary Medical Association Journal* (vol. 22, nos. 2 and 3, July–December 2003: 47–48).

*Minimum Vaccine Protocol for Cats*

- When twelve weeks or older, give FCV (calici), FVR (herpes/rhino), FPV (panleukopenia), and then rabies, but only if required by law. (It is good to give the rabies vaccination separately, three to four weeks later.) PureVac, canary pox vectored rabies vaccine (Merial), is preferred for cats. Vaccinating against *Giardia* is not advised since the vaccine can cause granulomas (masses of inflammatory cells that could become malignant).

- FeLV (feline leukemia) vaccine should only be given to at-risk cats at nine and twelve weeks or twelve and fifteen weeks with a booster at a year of age and none thereafter, in order to reduce the chances of injection-site fibrosarcoma, a cancer that can be fatal.

- Cats should have serum titer tests for FPV later in life to determine their immune status. All vaccinations to be injected under the skin should be placed as far down the cat's limbs as possible, since it is more difficult to treat fibrosarcomas that develop at other sites, such as the neck and back. This injection-site protocol is important since a cat can be saved from injection-site cancer by leg amputation, but has little chance of survival if the cancer develops between the shoulder blades or nape of the neck, where some veterinarians are still injecting vaccines.

No vaccine can guarantee immunity, since different strains of infective agents may be involved, and animals that are stressed or suffering from poor nutrition, genetic susceptibility, or concurrent disease may have impaired immune systems and lowered resistance to disease. But this does not mean that they should never be vaccinated or be routinely revaccinated just in case, because vaccinations can cause further immune system impairment and a host of health problems (the so-called vaccinosis diseases) that these new vaccination protocols are aimed at minimizing.

## Prevention or Profit?

The love that is shared between most people and their animal companions makes both parties vulnerable to the exploitation of those whose vested interests in making a profit from selling some new product or service is only too often frivolous and contrary to the best interests of the animals and the animals' care-givers. Those who resort to a kind of emotional blackmail (saying, If you don't give this new pet food or disease-and-suffering-preventing product to your pet, you are an uncaring and irresponsible person) operate in a money-driven ethical vacuum in total disregard of the real costs and consequences.

In the final analysis, we are beginning to make amends for our limited appreciation and understanding of what wellness in body, mind, and spirit means. With a deeper understanding of the health and behavioral problems that afflict our companion animals—which mirror, in so many ways, our own afflictions and diseases—we may find the way to wholeness, health, and fulfillment.

# 16
## | Preventing Fleas, Ticks, and Mosquitoes Naturally |

This topic is important to me because of the adverse reactions many animals have to the new anti-flea and tick medicines, the environmental risks of these chemicals, the suffering of animals allergic to fleas, and the increasing risks of tick-borne and other insect-transmitted diseases to companion animals, to wildlife, and to us—in part associated with climate-change/global warming. Since these chemicals (and also the heartworm preventive medicine, ivermectin) are excreted in treated animals' stools, fecal material should not be left in the open or flushed down the toilet, but should be bagged and put in with separated, biodegradable household garbage to go to a hopefully well-contained and managed municipal landfill.

The holistic approach to flea and tick control detailed below helps reduce the need to give cats and dogs potentially harmful new anti-flea and tick medicines (as pills, spot/drops on the skin, sprays, dips, and collars). These new medicines do not actually eliminate ticks and fleas, and when there are many, the additional control measures detailed below must be adopted anyway. Any and all control measures are jeopardized when cats and dogs are allowed out to roam free in the neighborhood. Collars are especially risky—to the cat and anyone sitting close by and petting the animal, especially children—since the chemicals are inhaled as well as absorbed by the animals.

These systemic insecticides that variously kill and disrupt the development of these and other parasites have to be ingested by ticks and fleas for them to work. This means that they must have at least one meal of the animal's blood before getting the poisons in the medicated pet's blood into their systems. Hence these new treatments do not stop infected ticks from transmitting Lyme disease, as well as Ehrlichiosis, Q-fever, Babesiosis, Tick paralysis, and Rocky Mountain Spotted Fever; they don't stop fleas from spreading the plague and Murine typhus; they don't stop fleas from causing allergic hot-spots in animals allergic to the insects' saliva. Mosquitoes can cause a nasty allergic reaction in cats called eosinophilic granulomatosis, creating itchy, weeping sores.

Cold winter is the best control, but in warmer states where it never gets cold enough to kill off fleas, the convenience of using these new anti-flea and tick meds should be weighed against the risks to the animals. Any animal showing any adverse reaction—lethargy, nervousness/irritability, nausea, poor coordination, and more severe neurological symptoms—should be taken off the drug at once. Very young, aged, sick, and nursing animals may be especially at risk, as well as those that have been recently vaccinated, since these drugs could compound any adverse effects of vaccinations on the immune and neuro-endocrine systems of companion animals.

## A Holistic Approach

My holistic approach to keeping fleas and ticks at bay consists of:

1.  Check daily with a flea comb.

2.  Closely examine between the animal's toes and ear-folds.

3.  Note any telltale shiny, black, coal-dust-like specks that turn reddish-brown on a piece of wet white paper, indicating flea droppings of digested blood.

4.  Fleas and unattached ticks caught in the comb can be quickly disposed of by dunking the comb in a bowl of warm, soap-sudsy water. Attached ticks should be removed by grasping the tick with tweezers as close to where it is attached, using a straight pull; twisting will break off the neck of the tick and leave its head buried in the animal's skin.

5. Vacuum all areas where the animal goes in the house every week, thoroughly.

6. Put cotton sheets over favored lying areas, such as sofas, carpets, and floor surfaces with deep cracks or crevices where flea larvae can hide and mature.

7. Roll up and launder these sheets in hot water every week.

For an already infested house, use insecticidal aerosols or "foggers" following all operator instructions, or call in a professional exterminator and put your animals in a boarding facility or motel during home-extermination *only* after each has been treated with a relatively safe pyrethrin-based anti-flea shampoo, or with an emulsion of neem and karanja oil rubbed into the fur. A second round of fogging the house and shampooing/dipping the animals may be needed, since flea pupae developing in cracks and crevices in the house may not be killed during the first treatment and may subsequently hatch out and start biting people and animals in the home. Dusting the animal with diatomaceous earth—a super-fine, harmless powder of fossilized microscopic sea creatures—purportedly kills fleas and their larvae by desiccation. (Birds often dust-bathe, probably to get rid of feather mites in this way.) Liberally sprinkling this same material (or borax powder) to act as a flea desiccant on floors, carpets, and in wall crevices, then vacuuming up after twenty-four to forty-eight hours, and repeating every two to three weeks during flea season, will help keep the home environment clear—provided animals living there do not roam free and come home infested. Some animals may not react well to this dust, which should be applied outdoors.

When control measures break down and fleas are found on the animal and cannot be kept at bay with regular flea-combing and other controls in the animal's environment, one of the safer flea-control products are those containing the oils and essences of chrysanthemum flowers, which paralyze fleas and are considered the least toxic to animals of all the insecticides, namely natural pyrethrins and synthetic pyrethriods. Repeated spraying, powdering, or shampooing is often needed since not all paralyzed fleas die on the first exposure.

Clean all porch, yard, patio, and garage areas of old mats, debris, brush, and dead vegetation where fleas and ticks may hide and flourish, especially in those areas where animals like to lie. Remove all old tires, plant pots, and other objects where rainwater may collect—including clearing blocked gutters and drain or fill areas where water pools—in order to control mosquitoes. Please avoid using ultraviolet light, electrocution bug-zappers, and spraying insecticides that kill millions of beneficial insects. Instead, put citronella candles out on the patio and in garden areas as repellants, put up insect screens on porches, and repair door and window screens.

A small lamp with a 20 or lower wattage bulb angled low over a large flat dish of soapy water or vegetable oil will become a heat-magnet and trap for hungry fleas in an empty house. This can be an alternative, when set up in different rooms, to fumigation while you are on vacation or purchasing a new home where there were animals.

Spritz your dog or cat daily with a floral-scented shampoo or hand soap, diluted in warm water; rub it into the fur and let it air-dry. This will change the scent signal of your companion animal and may help deter insect pests. Putting a few drops of oil of lemon and eucalyptus, neem and karanja, or cedar and peppermint (or try a mixture of various combinations of the same) in a cup of warm water, then shaking vigorously and spraying on the animal's fur, especially around the ear tips to also repel biting and flesh-eating flies, may significantly help repel fleas and ticks and mosquitoes. The lemon and eucalyptus oil combination has been recently approved for human use by the FDA as a safe and effective alternative to DEET to repel mosquitoes. But be prudent, especially with cats who should not be allowed to lick off these various sprays or hand-applied emulsions. Slicing a lemon, placing it in a cup of boiling water, and letting it stand overnight will provide a quick emergency potion that can be rubbed into a dog's fur and allowed to dry to repel fleas and other insects.

A bed for the animal stuffed with cedar shavings, mixed with crushed neem leaves and bark, and dried bunches of rosemary and lavendar may help deter fleas and keep them off an animal lying on such a bed. Few animals to my knowledge are allergic to these various plant materials. Pennyroyal has been advocated as an herb that helps repel fleas, but it has fallen into disuse because it can be toxic if ingested.

Healthy animals are less attractive, for reasons that science has yet to determine, to fleas and other external and internal pests and parasites whose whole existence is one of opportunistic survival and multiplication.

I have received many letters affirming this from readers of my "Animal Doctor" column. A classic example is one reader on the East Coast who wrote to say that her cat never gets fleas because she feeds her cat a natural home-prepared whole-food diet, and supplements his diet with brewer's yeast, while her neighbor's cat who also goes outdoors like hers (a practice that I deplore if it is not in the confinement of a backyard enclosure) always gets fleas in late summer.

So I advise giving brewer's yeast or nutritional yeast (not baker's or bread-making yeast), about one teaspoon per thirty pounds of body weight, mixed into the animal's food every day. This, like taking B-complex tablets, is the hunter's and fisherman's way of avoiding bug-bites (though I would wish that as nonsubsistence hunters they might find more compassion-filled modes of recreation). A teaspoon full of flaxseed oil (per thirty pounds of body weight) will also help improve skin and coat condition for both dogs and cats, though cats may do better on an organically certified fish oil. For most breeds of dogs, but not for cats, one daily garlic clove per forty pounds of body weight, chopped up and mixed into the food, may also help increase resistance or deterrence to fleas and other opportunistic bugs from the invertebrate world.

Fleas, ticks, mosquitoes, biting flies, and other insects we hate and fear are far more ancient than humans and their animal companions. We should try to keep these creatures at bay with the least harm to all. And that entails a holistic approach to animal health—a kind of ecological diplomacy based upon the ultimate empathy of enlightened self-interest that includes companion animal's emotional/psychological, as well as physical, well-being (the two being inseparable in making for a well functioning immune system).

The best medicine is prevention, and a holistic approach to companion animal health in this twenty-first century calls for a revision of vaccination protocols, of feeding highly processed commercial pet foods, and of over-medicating—especially with so-called preventive medications like those sold to keep fleas and ticks at bay, especially when there are safer and cheaper alternatives with far less risk to animals' health.

# 17
## | Endocrine-Immune
## Disruption Syndrome |

Chemical compounds called *endocrine disruptors* may play a significant role in various chronic diseases in both companion and other animals, and also in humans. These diseases include: allergies; chronic skin diseases; recurrent ear, urinary tract, and other infections; digestive system disorders such as chronic colitis, diarrhea, and inflammatory bowel disease frequently associated with immune system impairment; and metabolic and hormonal disturbances expressed in a variety of symptoms, from obesity to thyroid and other endocrine disorders, especially of the pancreas and adrenal glands.

Veterinarian Dr. Alfred J. Plechner's clinical findings that link elevated serum estrogen levels, thyroid dysfunction, and impaired synthesis of cortisol with a variety of health problems in animals warrant careful consideration, more detailed research, and randomized clinical trials. His claimed benefits of very low doses of cortisone, often in combination with thyroid hormone replacement, may hold true for some patients suffering from what I term the *Endocrine-Immune Disruption Syndrome* (EIDS). But long-term cortisone treatment may aggravate the syndrome, especially in the absence of a holistic approach to improving the animal's immune system and overall physical and psychological well-being.

Adverse reactions to vaccinations, anti-flea and tick medications, and other veterinary drugs, as well as hypersensitivity to various foods and dietary additives, may be consequential and contributory elements in what I interpret as a widespread and not yet well-recognized Endocrine-Immune Disruption Syndrome. I receive many letters from readers of my syndicated newspaper column concerning dogs and cats with the kinds of chronic, complex, multiple health problems that conventional veterinary treatments have at best only temporarily alleviated.

The primary cause of these hormonal imbalances and associated neuro-, endocrine, and immune system dysfunctions is most probably environmental in origin, specifically the endocrine-disrupting compounds (EDCs), such as dioxins, arsenic, PCBs, and various pesticides, in animals' food and water. Through bioaccumulation, these compounds become concentrated in various internal organs of companion animals, in farmed animals raised for human consumption (including aquatic species), and also in wildlife and humans at the top of the food chain. Many EDCs are *lipophilic*, meaning they especially accumulate in animals' fatty tissues, brains, mammary glands, and milk.

## Cats and Household Chemicals

Cats are extremely sensitive to environmental chemicals. They are like the proverbial canaries down in the coal mines, often warning us when they get sick that there may be something in the home environment that is harming them, and possibly us too. I have correspondence from cat owners who have used Swiffer (propylene glycol based), and Chlorox-Clean Up, and other floor and counter-top cleaners, whose cats became ill, and recovered after they stopped using these widely marketed products. Such products, and the marketing epidemic of antibiotic wipes and cleaners, that can only result in yet more resistant strains of potentially harmful bacteria, are to be avoided for our own good and for our cats' sake. Stick to the safe and tried and true cleaners, like white vinegar, ammonia, borax, and the new "green" cleaners, laundry detergents, organic orange/citrus cleaners. These are less harmful not only to our animal companions but to the environment and wildlife, as well as to ourselves. Most conventional cleaners and detergents, along with a host of cosmetics, toiletries, sunscreens, and other widely used products, are hormone-disrupting, endocrine-mimicking compounds that cause havoc to the human and other bodies. Cats walk constantly, with bare paws, over many surfaces in our homes that we have treated with potentially lethal chemicals, and lie on materials to which they may be allergic, and which we may have laundered with harmful chemicals.

Critics of the above statements will say that I am fostering chemophobia, but their days are sorely numbered as more and more people are doing their own research on the Internet. Perhaps that is why Swiffer no longer indicates on the container of its disposable pad floor cleaners what the chemicals are in the product, yet still warns to keep their product away from children and pets.

An Internet search and review of the existing literature and ongoing research in the field of environmental toxicology will reveal the ubiquitous presence of (EDCs) in the environment, especially from industrial pollutants (from power plants and municipal incinerators, to paper mills and chemically dependent industrial agriculture) and from untreated and inadequately treated sewage water (some 850 billion gallons of which are dumped annually into U.S. waters). EDCs are also being identified in a host of household and medical products, especially plastics, in clothing, floor materials, the lining of food cans (notably phthalates and Bisphenol A), and in the food and water we share with our companion animals and give to farmed animals. Female sex-hormone-mimicking phthalates, widely used by the food and beverage industry in plastic food wrappers and containers, have been recently implicated as contributing to obesity in men!

New EDCs are being identified, detected in human breast milk, infant umbilical cord blood, and in "signal" wildlife species, from alligators to Arctic seals. Researchers with the U.S. Geological Survey (USGS) Contamination Biology Program have found that PCB-treated fish have lower resting plasma cortical titers and disrupted stress responses, impaired immune responses, and reduced disease resistance. PCBs disrupt glucocorticoid responsiveness of neuronal cells involved in the negative feedback regulation of circulating cortical levels. I link these and other research findings on EDCs with Plechner's findings of low serum cortical levels in his patients, exposed undoubtedly to a number of EDCs that can have enhanced toxicity through synergism. Ironically the USGS has found human birth control estrogens in river waters.

EDCs not only disrupt endocrine signaling systems (estrogen, progesterone, thyroid, glucocorticoid, retinoid, etc.) and immune system functions, they can also cause profound behavioral, neurological, and developmental disturbances. They may play a role in obesity and in animals' adverse reactions to vaccines, other biologics, and pharmaceutical products.

There is an urgent need for the veterinary profession to address this Endocrine-Immune Disruption Syndrome, and to consider it when treating a variety of chronic diseases in animal patients. For a start, all veterinary practitioners should encourage animal care-givers to provide sick (and healthy animals, as part of holistic health maintenance) with pure water and organically certified food, including diets with animal fat and protein derived from young animals fed and raised organically and not exposed to herbicides, insecticides, and other agricultural chemicals, and veterinary pesticides and other drugs (even synthetic pyrethrins are powerful endocrine disruptors). Seafoods in the diet, especially of cats, should preclude species high on the food chain, like tuna and salmon. Also livestock that is organically certified should not be fed fish meal because of the bioaccumulation of EDCs. Many commercial dog and cat foods are high in soy/soya bean/vegetable protein. Since soy products are high in plant estrogens (those from genetically engineered soy being potentially extremely problematic in this matter), it would be advisable to take all animals suspected of suffering from EIDS off all foods containing phytoestrogen-laden plant proteins, and for healthy cats not to be fed any diet that relies on soy as the main source of protein. Healthy dogs, which are more omnivorous than cats (who are obligate carnivores), may not be at such risk.

The use of so-called xenobiotic detoxification enzyme and other therapeutic nutrient-supplement treatment, as detailed by Dr. Sherry A. Rogers and Dr. Roger V. Kendall (see References on the next page), is worth consideration for chronically ill animals that may have EIDS. These include essential fatty acids, as in flaxseed oil, digestive enzymes (e.g., papain and bromeliad), and vitamins A, B complex, C, and E, alpha-lipoic acid, L-carnitine, L-glutamine, taurine, glutathione, dimethylglycine, CoQ10, bioflavinoids, selenium, copper, magnesium, and zinc (with caution as per breed susceptibility to toxicity).

Homeopathic practitioners use nux vomica and sulfur to help detoxify a patient.

Detoxification can also include a bland, whole-food, natural diet for three to five days (individual food-hypersensitivity being considered), including steamed carrots, sweet potato and other vegetables; cooked barley or rolled oats; and a little organic chicken or egg, plus a sprinkling of kelp (powdered seaweed), alfalfa, or wheat grass sprouts, and milk thistle. For cats, the amount of animal protein should be at least two-thirds of the diet, while one-third is sufficient for dogs. After this cleansing diet, a whole-food, home-prepared balanced diet is advisable. In some cases, fasting for twenty-four hours may also be beneficial prior to giving the detox diet. But caution is called for since this could put some cats at risk.

The use of lawn and garden pesticides and other household chemicals, especially petroleum-based products, which could be endocrine disruptors, should be avoided, as well as plastic food and water containers for all family members, human and nonhuman. New carpets, plastic chew-toys, and stain-resistant fabrics and upholstery may also be potential hazards.

The medical and veterinary evidence of an emerging EIDS epidemic is arguably being suppressed for political-economic reasons: witness the U.S. government's foot-dragging from one administration after another to take effective action to phase out hazardous agricultural chemicals and industrial pollution to protect consumers from dioxins, PCBs, and PBBs—all potent EDCs. These compounds in particular contaminate through bioaccumulation the foods of animal origin—the discarded and condemned parts of which are recycled into pet foods and livestock feed.

*References*

Theo Colborn, et al., *Our Stolen Future* (updates at: www.ourstolenfuture.org/NewScience/newsources/newsrce.htm; see also: www.ourstolenfuture.org/New/recentimportant.htm).

Environmental Working Group, "Body Burden: The Pollution in Newborns," www.ewg.org/reports/bodyburden2.

R. V. Kendall, "Therapeutic Nutrition for the Cat, Dog and Horse," in A. M. Schoen and S. G. Wynn, eds., *Complementary and Alternative Veterinary Medicine* (St. Louis, Mo.: Mosby, 1997): 53–72.

Sheldon Krimsky, *Hormonal Chaos* (Baltimore: Johns Hopkins Press, 1999).

Alfred J. Plechner, *Endocrine-Immune Mechanisms in Animals and Human Health Implications* (Troutdale, Ore.: NewSage Press, 2003).

T. G. Pottinger, "Topic 4: 10 Interactions of endocrine-disrupting chemicals with stress responses in wildlife," *Pure and Applied Chemistry* (vol. 75, 2003): 2321–2333.

S. A. Rogers, "Environmental Medicine for Veterinary Practitioners" in A. M. Schoen and S. G. Wynn, eds., *Complementary and Alternative Veterinary Medicine* (St. Louis, Mo.: Mosby, 1997): 537–560.

U.S. Department of the Interior, U.S. Geological Survey, "Summary of Endocrine Disruption Research in Contamination Biology Program," updated 11-10-04, http://www.cerc.usgs.gov.

# 18
# | Pure Water for Cats and Dogs—and All |

It is difficult to find pure water almost anywhere on the planet, because of chemical contamination. This contamination stems from pesticides, heavy metals like lead and mercury, copper from pipes, arsenic compounds, radioactivity in some regions, excessive amounts of nitrates and phosphates, potentially harmful bacteria and other microorganisms, even pharmaceutical products excreted by humans and other animals given various drugs, and also industrial pollutants, especially dioxins and PCBs. Pollution of the air means contaminated rain, and polluted lakes and oceans mean contaminated rain through the hydrological cycle of evaporation and poison-cloud formation.

Water-treatment facilities and most water-purification systems (like reverse osmosis, ultraviolet and ozone disinfection, ion exchange, and activated carbon filters) do not get rid of all of these contaminants, which can pose serious heath problems to us and to our animal companions. The widespread chlorination of water to kill bacteria causes further problems, especially when there are high levels of naturally occurring organic contaminants because byproducts like chloroform and trihalomethanes are formed that are highly carcinogenic—causing kidney, liver, and intestinal tumors, and also kidney, liver, and brain damage, as well as birth and developmental defects in test animals. Alternative water disinfection with chloramines, a hoped-for safer alternative to chlorine, also results in the formation of highly toxic iodoacetic and holoacetic acids.

## Fluoride and Other Problems in Water

Compounding the health hazards of water treatment of already contaminated water is the addition of fluoride, a byproduct of the phosphate fertilizer industry, ostensibly to strengthen people's teeth and prevent cavities. But in order to do this, fluoride must be in direct contact with the teeth. That means the fluoride must be applied topically. Studies have shown no benefit from ingested fluoride.

On the contrary, fluoride can mottle the teeth and cause a host of health problems—notably osteoporosis, arthritis, kidney disease, and hypothyroidism. Fluoride has also been linked with gastrointestinal ailments, allergic skin reactions, impaired cognitive ability in children, harm to the pineal gland that helps regulate the onset of puberty, and possibly cause cancer. Flouride in tap water has been linked with osteosarcoma, particularly in young boys.

Of particular concern is where there is some already existing kidney disease, the kidneys' ability to excrete fluoride becomes markedly impaired, leading to a build-up of fluoride in the body.

I am also very concerned about the widespread use of aluminum chlorhydrate and other aluminum compounds, implicated by some to play a role in Alzheimer's disease, and various anionic and cationic emulsion and powder polymers, especially polyacrylamide, for waste-water treatment. Used to cause flocculation and coagulation of various wastes, including the effluent from poultry slaughter/packing plants, some wastes that contain these added agents may be variously used as fertilizer and livestock feed. Acrylamides are carcinogenic and can cause genetic damage, neurological problems, and birth defects, and may therefore finish up in our food chain and drinking water. Flouride can combine with aluminium to form aluminium fluoride, which is implicated in Alzheimer's disease and can interfere with many hormonal and neurochemical signals. Clearly, trading one or more health risks from contaminated water for others created by water-treatment processes is ill advised. Safer, cost-effective, organic, microbial, and ecologically based alternative systems of water treatment, recovery, purification, and waste management/disposal are urgently called for.

It is for the above reasons that I advise all people to drink pure spring water (that usually only requires sand-filtration to meet with U.S. National Sanitation Standards certification) *and to provide the same for their animal companions*—particularly those who, for various health reasons, are drinking more water and who have impaired kidney function.

The mineral content of most spring water, and water from remote, often high altitude, glacial and other isolated sources far from industrial and agricultural activities, are generally extremely beneficial; but excesses need to be closely monitored because of possible trace-nutrient imbalances that they may cause, and also urinary calculi or stones. Purified, distilled water, lacking in these essential minerals, may actually cause osteoporosis and other health problems.

## The Importance of Quality Water for Cats

Some cats have a clear aversion to drinking tap water, a possibly natural instinctual reaction to potentially harmful chlorine, fluoride, and other contaminants. Many prefer to drink from a dripping faucet or cat water dispenser, possibly because there is less aversive smell than there is in standing water in a bowl.

Cats' aversion to tap water is compounded by many becoming addicted to dry food, which is associated with several health problems, especially inflammation of the urinary bladder and the development of stones/sand/urinary calculi, and urethral blockage in male cats (i.e., the so-called feline urologic syndrome). Cats, being of a desert origin and physiology, lack the normal thirst mechanism when their diet is dry and deprives them of fluids. So they may fail to properly regulate their fluid balance by drinking more to compensate for an all-dry-food diet and may suffer the consequences of the feline urologic syndrome and ultimately fatal kidney failure, which is now taking many cats at an early age.

Since cats drink little water at the best of times, therefore, they should always be given pure spring water, or the closest equivalent, since poor quality, contaminated water will only worsen their already compromised condition of inadequate hydration.

Older animals, like older people, at risk from chronic heart and kidney disease should not be given water that has been treated with salt (sodium chloride) to soften it. Hard water for domestic use is often treated with salt to soften it; especially in apartments and condominiums, treatment of the central water supply to both cold and hot water faucets is not separated, which means that soft/salt-contaminated water comes out of both faucets. Older animals with chronic kidney disease, diabetes insipidus, and other health problems, and those on certain medications like prednisone, will drink copious quantities of water—and will thus be more at risk from absorbing more than normal quantities of waterborne chemical contaminants.

Humans have disrupted and poisoned the hydrological cycle/system on this "blue water" planet Earth, at great cost to our own health and to all other life forms that, like us, need water to live. Many live in the polluted surface waters of the planet from which they have no escape—nor ultimately, do we.

We are drinking the poisons that we thought nature—the ecology—could somehow assimilate, dilute, and neutralize. But that is not the case, as any chemical analysis of human and whale mother's milk will affirm. But there are long-term solutions beyond the science and technology of water purification and desalinization, which alone cannot guarantee a safe and sustainable source of drinking water for the generations to come. These include not using pesticides (herbicides, insecticides, fungicides, and chemical fertilizers) on our lawns, gardens, golf courses, and crops—making organic and sustainable agriculture one hope for the future; deconstructing the waste-disposal, incineration, energy, petrochemical, paper, plastic, and other consumer-driven commodity and appliance industries, so that the fewer pollutants they release into the air and surface waters (and ultimately into our food and water chains), the more they profit; and in the process, market no products or byproducts that cannot be recycled without harm to the health, vitality, integrity, and beauty of this living Earth, and all who dwell therein.

## A Call to Action

We can no longer take water for granted as one of nature's bountiful, pure, and eternally renewable resources. Water is the fundamental life source that sustains all beings, and to our peril and the demise of this living planet, we have thoughtlessly squandered and poisoned this basic, vital element of existence. We cannot trust that the water coming from our faucets and wells is safe to drink or is safe to give to our animals—and the science of water safety and quality evaluation is still in its infancy.

Even so, consumer-citizen taxpayers have a right to have their municipal water authorities test domestic water sources regularly and make their findings available to the public. They also have the right to demand better monitoring and law enforcement to protect open waters from pollution runoff from people's lawns and gardens, as well as from agriculture and other human activities and various industries.

Just as the organic foods market has accelerated with increased, informed consumer demand over the past decade, so too in this next decade of the twenty-first century is a more informed public demanding pure water—some of the purest coming from ancient springs, more remote, higher-altitude, and glacial sources, and as-yet-uncontaminated and sustainable deep aquifers. But these sources will not last forever. It is the responsibility of all of us to conserve water, to stop polluting, and to treat this basic resource—the source that sustains all life—with respect and gratitude.

The northern states of the United States have an immediate and most urgent responsibility at this time to reduce the agricultural, industrial, and municipal sewage pollution of rivers and overexploitation of the same, through diversion and dam-construction for commodity-crop irrigation and dairy and other livestock production. (This has been notably facilitated for decades by the U.S. Army Corps of Engineers, which has put the Florida Everglades on the never-glades path to extinction for the sugar and cattle industries and for industrial-scale orchard/plantation irrigation after natural ecosystems have been annihilated—along with the many wildlife species like the wolf, the lynx, the Florida panther, the flying squirrel, and the coati-mundi.) The rivers/waterways that flow south across the North American continent harm all southern states that take what poisoned water is left, which ultimately flows to the Gulf of Mexico—where an area of ocean the size of Rhode Island is now void of any life, according to marine biologists and the local fishing industry.

All those who care about their health, the health of their families, and the health of their companion animals will see the wisdom of purchasing the best-quality water they can and stop adhering to the erroneous belief that their tap water is necessarily safe for their consumption. Water quality and safety is a wake-up call for us all, and the state of its quality confirms the truism that when we harm the Earth, we harm ourselves. The health of the Earth—of aquatic as well as terrestrial ecosystems, just like the health of the human population—is interdependent and linked by air and water quality that, for the good of all, we must improve and maintain.

# 19
## | The Right Diet |

### Commercial Nutritional Horrors

I believe if consumers were aware of the ingredients in many commercial pet foods, they would be outraged. Most contain condemned parts of factory-farmed animals that are considered unfit for human consumption. One would not use them as fertilizer in one's garden, let alone feed them to one's beloved pet.

In *Foods Pets Die For: Shocking Facts About Pet Food*[1] author Ann Martin presents more disturbing details and documentation on this issue. As Martin emphasizes, many ingredients are essentially leftovers from the human food and beverage industries, and they are lacking in essential nutrients, which manufacturers compensate for by adding synthetic additives and supplements. Processing at high temperatures destroys many so-called heat-labile nutrients. Preservatives and added stabilizers, flavor enhancers, and coloring agents, especially in dry and semi-moist cat foods and snack-treats, make for a chemical soup that can lead to chemical sensitivity and food allergies.

After endorsing the first edition of Martin's wonderful book in 1997, I was actually put on disciplinary probation (and essentially silenced) and my salary was frozen until my mandated retirement from the Humane Society of the United States in 2002, by the then-president of HSUS. The pet food industry had sent a letter of complaint to my boss, who was hoping to secure funds from pet food companies!

---

1. Ann N. Martin, *Foods Pets Die For: Shocking Facts About Pet Food* (Troutdale, Ore.: NewSage Press, 1997).

*New Hope for Companion Nutrition*[2]
There is a niche market that is expanding now for organically certified pet (and human) food. There are several excellent books by veterinarians and other informed animal nutritionists on home cooking for animal companions. They present recipes for making nutritious and well balanced foods from scratch for dogs and cats (see Ann Martin's book for recipes, as well as my own basic recipe included in this chapter).

This new trend is by far, I believe, the best way for all conscientious consumers to better provide healthy nutrition for their animal companions. Alternatively, one can purchase organic, high-quality commercial pet foods that are based on sound nutritional science. You can also feed a combination of both homemade, whole foods and organic commercial kibble. Avoid commercial food with an ingredients list stacked early on with corn meal. Such foods are too high in starches and too low in essential fatty acids.

## A Home-Cooked Meal

Home-prepared foods for our animal companions, ideally with organic ingredients that were locally produced, are important because you then know what your animal is being fed, if a food-related health problem such as an allergy to a particular ingredient or digestive upset were to arise. Most processed commercial pet foods contain all kinds of human food–industry byproducts and ingredients that are considered unfit and unsafe for human consumption. Many are of questionable nutritional value after repeated processing. In short, you just don't know what you're feeding your cat. Aside from coloring agents that may cause problems other than saliva-staining of animals' faces and paws, most commercial pet foods contain: artificial preservatives like BHA, linked with cancer of the bladder and stomach; BHT, which may cause cancer of the bladder and thyroid gland; and Ethoxyquin, one of Monsanto's many allegedly harmful products that renderers (meat and poultry processors) add to the fat/tallow put into pet foods to prevent rancidity. Ethoxyquin is a recognized hazardous chemical and a highly toxic pesticide.

2. For clinical documentation of the dangers of high cereal content dry cat foods, see *Your Cat: Simple New Secrets for a Longer, Stronger Life* by Elizabeth M. Hodgkins, DMV, Esq. (New York, N.Y.: St. Martin's Press, 2007).

*Dr. Michael Fox's Homemade "Natural" Diet for Cats*
Below is my recipe for a homemade diet for your cat.

> 1/2 cup uncooked brown rice
> 1/2 cup peas, chickpeas, or lentils
> pinch of salt
> 1 tablespoon vegetable oil (flaxseed oil or safflower oil)
> 1 tablespoon wheat germ
> 1 tablespoon cider vinegar
> 1 teaspoon chopped canned clams in juice
> 1 teaspoon nutritional yeast
> 1 teaspoon dried kelp
> 1 teaspoon bone meal or calcium carbonate
> 1 whole chicken cut into pieces, or 1 lb. hamburger (not
>     too lean), ground lamb, or turkey

Combine all above ingredients except oil. Add water to cover ingredients, simmer for 35-45 minutes, and stir; add more water as needed until cooked and thickened. Stew should be thickened enough to be molded into medium-size or muffin-size patties (add bran to thicken if needed). Also add an egg or cup of cottage cheese. Immediately after cooking and cooling, debone and discard bones (cats should not be given cooked bones to eat since they can splinter and cause internal damage). Once cooled, add oil.

This stew can be served, in the amount of one-half cup per day, to your cat with the rest of his/her rations. Freeze the rest of the stew as patties, or in muffin trays, and thaw out as needed. Serve one patty to a cat about three times per week with regular rations.

For variation, substitute 1 pound raw or lightly cooked boneless fish. (Note: Some cats are allergic to fish, corn, and also to beef and dairy products.)

*These items are available in health food stores. Ideally all ingredients should be organically certified.

The above recipe can also be fed as a complete meal rather than as a supplement. Mix increasing amounts of your cat's new food with decreasing amounts of the old food over a seven-day period to avoid possible digestive upset. It is advisable to vary the basic ingredients to provide variety and to avoid possible nutritional imbalances, and to monitor the animal's body condition so as to avoid either overfeeding or underfeeding, based on the average cat consuming one-third of a cupful three or four times a day. Note that different animals have slightly different nutritional needs according to age, temperament, amount of physical activity, and health status. Reduce the portion size if you are feeding your cat more frequently. Many cats enjoy small meals six to eight times a day, with a very small amount of dry food, or home-baked kibble given as a daily treat and possible tooth-cleaner.

## Pet Owners Get a Shocking Wake-Up Call Involving the Commercial Pet Food Industry

In March 2007 millions of concerned pet owners became aware of the massive recall by Menu Foods (a pet food manufacturer in Canada) of sixty million cans and packages of contaminated, poisonous cat and dog food. The recall was an effort to prevent the development of acute kidney disease and even death in the nation's pets.

This recall eventually involved several thousand varieties of cat and dog food and about a hundred different brand names and distributors, including major well-known ones such as Iams, Eukanuba, Nutro, Hills, Nutriplan, Royal Canin, Pet Pride, Natural Life vegetarian dog food, Your Pet, America's Choice-Preferred Pet, Sunshine Mills; store brands such as PetSmart, Publix, Winn-Dixie, Stop and Shop Companion, Price Chopper, Laura Lynn, KMart, Longs Drug Stores Corp., State Bros. Markets, and Wal-Mart; and a host of private labels of mainly canned (moist) cat and dog foods. When coupled with the soon-to-follow recalls of other pet food manufacturers that did not contract with Menu Foods, notably other well-known company brand names like Purina, Alpo, and Del Monte Pet Products, the quantity of food recalled would be in the hundreds of thousands of tons.

I began to receive letters from dog and cat owners thanking me for "saving their animal's lives," because they were feeding their pets the kind of homemade diet that I have been advocating as a veterinarian for some years. Other letters document the suffering and death of several companion animals, their caregivers' disbelief and outrage, and their financial as well as emotional losses. Many people had veterinary bills in the three- to six-thousand-dollar range, and those on fixed incomes had to take out credit card loans and pay exorbitant interest. I received some letters that described animals going into sudden acute renal failure prior to the purported November onset of this tragedy. Others reported liver damage.

*Scientists Investigate What Is Killing Pets*

Toxicologists eventually identified two chemicals that had been put into imported wheat flour by two fraudulent Chinese manufacturers, who sold the contaminated product as wheat gluten and rice protein. These chemicals were melamine and cyanuric acid, both related to urea, which caused crystals to form in animals' kidneys. But not all toxicologists were in agreement that these chemicals, which are put in small amounts into livestock feed, were the only culprits. Some ingredients could have come from genetically engineered crops and be laced with kidney-harming herbicides, and acetaminophen was also discovered in some pet foods, which can cause liver damage, especially in cats.

On April 30, 2007, the *Washington Post* quoted a Chinese food and feed processor as asserting that for several years urea was added as cheap filler, and that it would pass as protein under the crude ingredient and quality tests done on food commodities by other countries. But when processors mixed too much urea into the vegetable protein/gluten livestock feed, it made animals sick, so they switched to melamine. Because this adulteration has been going on for years, it is quite probable that importers of vegetable protein from China have been putting U.S. consumers and their animal companions at risk for some time.

The possibility of synergism of toxic pet food contaminants—where two or more harmful additives or contaminants and/or their metabolized breakdown products result in pet food poisoning, as this pandemic demonstrates—still remains open.

When challenged by Senator Richard Durbin at Senate hearings in April, representatives for the Pet Food Institute and the American Association of Feed Control Officials—whose AAFCO labeling, though standard on most processed pet foods, gives no valid guarantees on quality or safety—became extremely defensive and contradicted themselves when it came to actual inspection and testing of ingredients.

The testimony of Dr. Elizabeth Hodgkins, a feline specialist in private practice in California who had served as Director of Technical Affairs at Hill's Pet Food Nutrition, was the one clear voice of reason and truth. Her testimony cut through all the cross talk and obfuscation to document the pet food industry's lack of effective regulation, oversight, quality and safety controls, adequacy and accuracy of labeling, and verifiability of claimed nutritional value. She made it quite clear that cats especially were becoming ill and dying well before the melamine scare because of some of the ingredients and formulations U.S. pet food manufacturers were marketing to a trusting public—and which far too many veterinarians still believe are scientifically formulated, balanced, and in accord with the claims on the AAFCO standardized label.

*Pet Owners Learn the Limits of the FDA's Powers*
The FDA has no mandatory authority to demand a pet food recall. All recalls are "voluntary," upon written request notification by the FDA. There is no mandatory requirement for pet food manufacturers to inform the FDA in a timely fashion, nor is there any penalty for not doing so.

This current debacle of the commercial processed pet food industry puts us all on notice. Better quality controls, oversight, and testing are called for, but one must be realistic. There have been recent massive recalls of human food commodities, including ground beef, poultry, onions, and spinach. Costs aside, no system of mass production can be fail-safe. The recycling of human food industry by-products and products considered unfit for human consumption into livestock feed and processed pet food presents a monumental risk-management challenge.

Why put wheat gluten and vegetable protein into cat food, except as a cheap filler that equates with poor nutrition for obligate carnivores like the cat? One legitimate exporter of wheat gluten in China said that the low market price for the fake wheat gluten that killed thousands of pets should have been suspect to the U.S. importers. But the bottom line is profit margins based on lowest-cost ingredients and feed formulations.

Perhaps, when all the numbers are in and the mortalities and morbidity rates of cats and dogs are more fully collated, the West should thank China for exposing, tragically as it was through the suffering and deaths of uncounted numbers of beloved animal companions, that our global food supply, in terms of quality and safety/healthfulness, is critically dysfunctional. Every government should, therefore, be called to address the deeper layers of the problem, of which the largest pet food recall so far is but the tip of the iceberg. Let us hope that there will not be many more such recalls. Appropriate government action, multilaterally, is our best hope.

In the interim, we should purchase produce for our families—and for many of us that includes our companion dogs and cats—that, ideally, is grown locally or in a known country of origin and is organically certified. Then the long-overdue revolution in agriculture and ethics may begin!

When will the pet food industry answer my question concerning the safety and nutritional value of genetically engineered food and livestock-feed industry products and by-products that are recycled into pet foods without any published animal safety tests? With such an expanding clinical caseload of allergies, skin problems, obesity, diabetes, immune disorders, and inflammatory bowel diseases, more and more veterinarians like me are advocating feeding companion animals whole foods that are ideally organically certified and prepared at home or are from a company that uses primarily organically certified, home-grown and home-derived ingredients. Many such ailing animals do far better on such diets than on the prescription diets that are promoted by what some call a brainwashed and ignorant veterinary profession. It seems as though the mainstream pet food industry is on a collision course with reality, still marketing mainly cereal-based dry kibble to carnivorous cats whose nutritional wisdom cannot be exercised when they have no choice or when they become addicted to the manufactured dry food.

## Cats' Special Dietary Needs

Cats require two essential amino acids in their diets: arginine and taurine. Vegetables and cereals, such as barley, oats, and rice, are deficient in these two essential nutrients. That's why it is not prudent to feed your cats only a vegetarian diet. *Cats are carnivores.*

A cat's diet must therefore include some form of good-quality animal protein that is easily digestible in order to supply these two amino acids. Cooked eggs, meat, fish, and dairy products provide these essential nutrients. Some cats are allergic to fish, and cats will develop other deficiency diseases if they are fed exclusively tuna fish, lean beef, or chicken. Many cats thrive on a bit of chopped raw meat, poultry, liver, or kidney (which should be first scalded or parboiled to kill off potentially harmful bacteria) added to their food.

Arginine helps rid the adult cat's body of ammonia, which is a breakdown product of digested protein. Arginine deficiency is associated with muscular tremors, incoordination, depression, and even death.

A deficiency in taurine eventually leads to blindness and can cause heart enlargement and weakness.

Some cat foods are supplemented with additional taurine and arginine. Check the labels. Remember, cats are *obligate* carnivores and require a high-protein diet, unlike more omnivorous dogs and humans, who can thrive on a balanced vegetarian diet. Cats also need a few drops of fish oil in their daily meals to provide them with essential fatty acids.

Cats should be given several (four to eight) very small meals during the day and evening since they naturally eat several small meals in the wild. No cat should be fed an entirely dry commercial cat food because they have a poor thirst mechanism—being originally a desert species with a high thirst threshold—and they may not consume sufficient fluids if their food is not moist. Plus, the high cereal content of most dry foods can be fatal.

It is advisable to get cats used to having their teeth brushed from an early age, since dental problems—build-up of tartar or scale, and gingivitis—are extremely common. Providing cats with the end portions of raw chicken wings to chew on, scalded in boiling water first to kill potentially harmful bacteria, will help keep teeth clean and gums healthy.

## Foods That Spell Trouble for Felines

Some cats, like those who suffer from diabetes, obesity, heart, liver, or kidney disease, need special diets that your veterinarian can prescribe. (Had they been fed diets high in meat rather than cereal-based cat foods, they would not need to be put on prescription diets and various drugs.) Be on the alert for food allergies, since some cats are allergic to cat food ingredients such as corn, soy, fish, beef, and dairy products.

Cardinal signs of food hypersensitivity include scratching, especially around the head and neck, patches of baldness, and the development of tell-tale papules or scabby lumps on the cat's skin. The cat vomiting soon after eating can mean he's allergic to certain foods. Some cats develop chronic diarrhea and inflammatory bowel disease.

Be alert to feline food allergy and avoid the temptation of dismissing signs as a skin disorder and putting salve on the cat's sores. The problem warrants veterinary advice, since chronic food hypersensitivity and nutritional deficiencies cause distress and weaken the cat's resistance to disease. The veterinarian can test the cat on a hypoallergenic diet of home-cooked rice and lamb or mackerel.

Many cats with food hypersensitivity begin to clear up after two to three weeks on this test diet, after which a vet will prescribe a special cat diet.

Problems arise when people get into the habit of giving their cats more than a taste of various human foods. Giving your cat just a whole can of tuna each meal is a wholly incomplete and unbalanced diet. Vitamin E deficiency disease and mercury poisoning could develop. More than a quarter cup of milk can cause diarrhea, because most cats are lactose intolerant. Too much milk is especially harmful for kittens.

Remember that meat or other high-quality protein is essential for cats, since certain amino acids that cats need are not available in vegetables, a deficiency of which can cause blindness, neurological problems, heart problems, and a general weakening of the cat's immune system. Likewise, raising kittens on an all-meat diet is as bad as trying to turn your cat into a vegetarian. An all-meat diet is unbalanced, high in phosphates, and low in calcium, which can result in bone disease and deformity.

Cooked eggs are a good animal protein source for cats, since heat deactivates the avidin in egg white, which would otherwise destroy an essential nutrient called biotin. All meat and poultry products and leftovers should be cooked and refrigerated in order to prevent bacterial food poisoning. Just like people, cats can suffer from salmonella and other bacterial diseases from contaminated meat, poultry, and seafood.

## Cystitis: A Common Cat Disease

If your cat suddenly begins to urinate on the sofa or on the floor, don't punish him or her too quickly. He may be suffering from cystitis and trying to let you know he is in pain. Cystitis, an inflammation of the bladder, is one of the most common feline diseases and is often chronic. Sand-like bladder stones and plugs of mucous frequently form and can block the urethra of male cats. The combination of cystitis and bladder stones is called feline urologic syndrome (FUS).

Warning signs of FUS include constant squatting and straining, blood in the urine, becoming unhousebroken, and blood or pus dripping from the vulva or penis. The cat may stop eating and drinking and become weak and dehydrated. You may be able to feel the cat's engorged bladder by touching the abdomen. If your cat shows any of these warning signs, take him immediately to the veterinarian for examination, or your cat could go into uremic shock and die.

Generally, a veterinarian will try to remove mucous plugs and bladder stones in the urethra with a catheter. Antibiotics may be prescribed for bacterial infections. In some cases surgery can enlarge a male's urinary tract so that it is less likely to get clogged.

Several factors seem to contribute to the development of FUS: possibly a virus, a diet high in magnesium, insufficient water intake, stress, and heredity. I advise that you wait until your male kitten is at least six months old before having him neutered. Early neutering results in a smaller-than-normal urethra, which is more likely to become clogged. Also, be sure your cat has plenty of water and don't feed it only dry food.

I recommend that you keep a cat with FUS off dry foods and give the cat a commercial prescription moist diet that your veterinarian can prescribe. Wean the cat onto a home-prepared diet like the one mentioned earlier (see pages 184–85). Since prevention is the best medicine, young cats should get used to regular moist canned or homemade food. Giving your cat a half-teaspoon of safflower or flaxseed oil in her food once a day may also help prevent cystitis.

*A Closer Look at Cystitis*

Uva ursi, or bearberry, can be a valuable anti-microbial and anti-inflammatory herbal treatment for cystitis, but it tends to make the urine alkaline, which is not desirable if a cat has struvite crystals in the urine. If crystals/urinary calculi are involved in the cystitis, it is important to determine what type they are chemically. With most types, acidifying the urine with Vitamin C or capsules of cranberry can help. But an overly acidified diet can lead to calcium oxalate crystal/urolith formation in the urine, so maintaining an optimal acid-base balance in the cat's diet, and normally slightly acidic urine, is essential. Encouraging the cat to drink more water, as by flavoring with milk, tuna juice, or unsalted beef or chicken broth, and not feeding an all-dry-type commercial cat food are additional steps to take. Diuretic herbs like parsley and dandelion help keep the urinary tract flushed out, but increasing the cat's fluid intake, as with flavored water and moist, canned or home-prepared foods, is important. Less cereal and more raw or lightly cooked meat to increase the acidity of the urine is advisable in many cases, unless they have calcium oxalate crystals in their urine.

Cats with cystitis and incontinence may have an underlying food allergy, and some have made spontaneous recoveries when all corn was eliminated from their diets—a common ingredient in many cat foods.

Cats with cystitis often need antibiotics because of underlying infection. Treatment with an analgesic like Butorphanol, or with Valium or reserpine, that have antispasmodic properties as well as being anxiety-alleviating, has helped many cats during the first five to ten days of treatment.

Diabetic cats and those on long-term steroids often develop bacterial cystitis because of lowered immunity. Incontinent cats should be checked for these problems.

Changing the type of litter to a non-mineral/clay base, like newspaper or wheat pellets, may also help.

Emotional stress, as well as an all-dry diet, is a major aspect of feline cystitis/urologic syndrome. Environmental enrichment, as with perches and climbing posts, a cat "condo," safe places to hide, identifying and resolving conflict between cats in the same home, regular play and grooming, providing an extra litter box, and putting out extra food and water bowls located in a quiet place were cats will not be startled or compete with each other are all helpful preventive measures.

# 20
## | Should All Cats Be Indoor Cats? |

Most dogs are natural roamers and hunters. But most people who love their dogs do not let them off their property; free roaming dogs could be hit by a car, injured in a dog fight, get caught in a trap in the country, picked up by an animal dealer and sold to an animal testing lab, get shot by a hunter, or get bitten by a rabid fox or raccoon.

In most municipalities across the United States—mainly because free roaming dogs might bite someone—there are laws that make it illegal to let one's dog freely roam the neighborhood. But this is not so for cats, and that is a serious problem since cats outnumber dogs in the United States by some 20%.

Too many cat owners, following family tradition, let the cat out at night and think nothing about allowing their cats to come and go from the house as they please, day and night. It's not that they love their cats less than the people who have dogs and take them for regular walks and romps in safe off-leash open spaces, never allowing them to roam off their property. And to suggest to such cat lovers that they are being irresponsible—and by implication, therefore, uncaring—is an affront. But free roaming cats get stolen, lost, hit by traffic, caught in traps, shot by hunters, even tortured by adolescents, and poisoned by bait put out to control rats; and many cats kill wildlife—from songbirds to squirrels and small rodents. When they eat small rodents they can become infected with toxoplasmosis, which is a major public health concern.

## Cats Outdoors: Health Risks to Cats and People

When free roaming cats defecate in people's gardens, which they often do, a pregnant woman could pick up toxoplasmosis and give birth to a brain damaged, visually impaired child. Similarly, a cat infested with round-worms could cause blindness and neurological and other problems—especially in children who develop visceral larva migrans from ingesting infective worm eggs in cats' stools, in placees such as playgrounds and sandboxes.

Feline AIDS, which does not affect humans, is a viral immunodeficiency disease that is spread via the saliva when an infected cat bites another cat. There is no cure, and the disease is generally fatal. Those people who let their cats roam free are on notice that they are putting their cats at risk from this disease, as well as other highly communicable cat diseases like feline herpes virus, feline calici virus, and feline panleukopenia. This latter disease in Florida has put the endangered Florida panther at risk, with many dying from this disease, which they contract from infected cats whose owners let them roam free. Bobcats, lynx, the gray fox, and other wild carnivores may also be at risk from infected house cats who spread this and other diseases in their bodily fluids.

Feline infectious anemia, or hemobartonellosis, can be transmitted by fleas from cat to cat and can be fatal if not quickly diagnosed and treated. Cats are especially prone to several highly contagious respiratory virus infections that can spread rapidly through a free roaming cat community—conditions that are often fatal for young cats or result in chronic infection and permanent impairment of the immune system.

Free roaming cats not only transmit diseases to each other and to wildlife, they can also bring diseases home to infect humans—especially children, the elderly, and others with impaired immune systems. They can infest the house with fleas that can carry plague, and with ticks that can carry Lyme disease and other serious tick-borne diseases like tick paralysis. Cats can get various diseases like Avian flu, Tularemia, Tuberculosis, Pneumocystis pneumonia, Campylobacteriosis, Cryptosporidiosis, Helico-bacteriosis, Giardiasis, Salmonellosis, Histoplasmosis, and Blastomycosis, usually from infected wildlife and contaminated soil and water, and then pass these diseases on to people. They often pick up ringworm from infected cats and pass this difficult-to-eradicate skin disease to anyone who pets them.

## Cats Outdoors: Further Ramifications

In spite of all these feline, human, and wildlife health concerns, cat owners still continue to let their cats wander off their property under the errone- ous belief that their cats need to roam and hunt, that it is their natural right, and that they instinctively know how to look out for themselves in the great outdoors.

But all of this is wishful thinking, and ultimately irresponsible and uncaring, especially when un-neutered cats become pregnant and owners then have the responsibility of finding homes for all the kittens—or as is usually the case, drop them off at the local animal shelter. Shelters are generally filled with unadopted kittens that only too often must be euthanized. Aside from the emotional burden on shelter staff, it is a tragic irony that those who claim to love their cats but let them freely roam and breed come up with various excuses. A common one is that it is educational for their children to see the miracles of birth and tender feline motherhood. A trip to the animal shelter would be far more educational.

Many litters of kittens are born in the wild and few survive. They usually die from exposure, starvation, and infectious viral diseases. Those that do survive quickly become feral, unadoptable wild cats that devastate songbird and other local wildlife species and compete with natural predators in suburban and rural areas across the United States. Cat owners do not see the magnitude of the problem when they let their one cat go out day and night, unaware that other uniformed cat owners are doing the same in their neighborhood—a place where there are probably many lost and feral cats carrying various diseases and ready to fight over territory, potential mates, and food. This is why one of the most common health problems requiring veterinary treatment in pet cats is cat bite abscess, often compounded by saliva-transmitted feline AIDS, both of which could be avoided by making cats enjoy life indoors and never allowing them to get a taste for roaming off the property.

The most common excuse for not making cats indoor cats is that they always fret and cry to be let out. Cats who have been raised indoors from kittenhood and have never been allowed out do not usually ever cry to be let out. Some who have not been neutered may cry and try to get out when they are in heat, or are being "called" by free roaming cats outside. Spaying female cats and neutering male cats at five to six months of age and younger can greatly help them adapt to life indoors.

Free roaming cats, unbeknownst to their owners, can become a menace to indoor cats when they prowl and yowl and spray around the house and threaten the residents through windows and screen doors. Some indoor cats become paranoid and more aggressive, attacking each other and their owners, and can start to spray inside the house and become unhousebroken.

There is nothing wrong with letting the cat out into a cat-proof backyard with a fence and a top wire-mesh lean-in panel that they cannot climb over. An alternative is to put up an outdoor "cat house," such as a chicken-wire-covered gazebo or A-frame structure with an insulated nest box or den for shelter and shade, a covered litter box, and a cat gym or outdoor condo where they can climb, scratch, and sunbathe. Alternatively, a screened-in porch or balcony can be similarly enriched with a secure tree branch or cat condo; cats can be provided additional stimulation by having a bird bath and feeder set up outside in clear view.

In most homes, people just have one cat, but cats are generally healthier and happier, and more entertaining, when they have another compatible cat as a housemate. Studies have shown that neutered cats get on best if they are littermates, or if you have a mother and one or more of her offspring. More experienced cat owners can introduce an unrelated kitten or young cat to provide social and emotional enrichment for a solitary cat who is not too old and set in her ways. Also, kittens have a critical period early in life between three and nine weeks of age during which time they most readily bond with humans and with each other. Kittens not handled until fourteen weeks of age and older are more fearful and aggressive toward people.

Cats are remarkably intelligent, perceptive, affectionate, and sensitive animals. And they can suffer from various forms of frustration and stress in the home and related emotional distress, which can lead to their abandonment or surrender to the animal shelter, where they will most likely be euthanized. The most common reasons for cats being put up for adoption—or for an as-humane-as-possible end to their lives at the animal shelter—are: destructive clawing of furniture and other household décor; aggression, toward other cats in the home or toward people; restlessness and loud calling/meowing, especially through the night and in the early hours; house soiling and not using the litter box all the time; constant hunger, food-solicitation, and vomiting; increased irritability, excessive grooming, self-mutilation, and unsightly alopecia. House soiling is extremely common and is one reason why some landlords and condo associations do not allow people to keep cats, for fear of the stink of cat urine permeating the property. Many of these problems are compounded by cats being allowed to go outdoors, and by declawing.

The reasons for the above behavioral problems are often medical, which means they can be treated, or psychological, which also means that they can be prevented or cured. Far too many cats are being abandoned and killed by owners who do not know, or do not care to know, that professional help is available in most communities to correct these problems that are not simply the result of cats becoming undisciplined or manifesting some irreversible personality change.

Through my nationally syndicated "Animal Doctor" newspaper column, I receive letters every week from people who are ready to get rid of the cats they love because they have one of the above problems—and have not sought or found appropriate professional help. In closing, I will list some of these problems for which too many cats are being punished, being physically and emotionally abused, are put outdoors under the erroneous belief that that is what they want, or are abandoned and even euthanized. These medical and psychological problems include anxiety, senile dementia, painful chronic arthritis, hyperactive thyroid disease, food allergy and inflammatory bowel disease, cystitis and related urological problems, and diabetes.

All cat owners should be prepared to face such medical and behavioral issues in their feline companions and have the financial and emotional commitment and understanding to see their cats through to the resolution of the problem, which most often did not mean that the animal had to be killed. The best medicine is prevention, which includes proper nutrition, and a cat's emotional and physical health and quality of life can be best guaranteed by providing a safe home environment that gives security, social stimulation, affection (as with another feline companion), and toys to play games with one or more humans in the family. Daily petting, grooming (see Chapter 21), and massage (as per my book *The Healing Touch for Cats*) are no less important than good nutrition and regular wellness examinations by the veterinarian. Owners of outdoor-indoor cats will find themselves facing more veterinary expenses for a variety of health problems, and they will be putting their cats' health at risk from additional vaccinations and anti-parasite and flea medications that indoor cats rarely, if ever, need.

# 21
## | Grooming Cats |

## A Guide to Grooming Your Cat

The act of grooming a cat causes deep relaxation and a dramatic slowing of the heart rate, thus helping the animal cope better with physical and emotional stresses. For aged and convalescing animals, grooming stimulates the circulation and improves their health. Regular grooming helps cement the cat-human bond and is also a good way to help prevent fur balls from accumulating in the cat's stomach from swallowed fur that the cat ingests during self-grooming. All cats need a daily grooming, and long-haired cats can do with a twice-a-day session to prevent matting and fur balls.

Animals should find grooming pleasurable. Some people find it's difficult to get their cats to submit to regular grooming. Perhaps they don't know how their animals like to be brushed. These guidelines will help to make weekly grooming a no-fuss affair.

- Get cats used to grooming early in life. Long-haired breeds especially should be groomed from the time they are kittens. Do not use tranquilizers on a routine basis to have thick mats of fur removed, since these drugs can lead to liver damage.

- Be sensitive to your cat's likes and dislikes. Cats often like to flop over onto one side for grooming, so don't insist that they sit up or stand. Work on that side first; then gently grab hold of their front and hind legs to roll them over to do the other side. Some cats don't like to be groomed near the base of the tail or along the abdomen, so be particularly careful with these areas.

- Use appropriate combs and brushes. For cats I prefer a fine, stiff-bristle brush. If the atmosphere is very dry and full of static, as it often is in winter, I moisten both my fingers and the brush, to reduce the chances of giving the animal electric shocks. For the same reason, place the animal on a wool rug or cotton towel rather than on synthetic nylon material.

- Be gentle and reassuring. Before you begin grooming, stroke your cat reassuringly around the head to make him feel safe and secure and to let him know that nothing unusual is going to happen. Also stroke the cat any time you accidentally knock a knee or shoulder with the edge of the brush.

- Learn to use the brush and your fingers properly. Begin grooming by running your fingers down the cat's back quite firmly several times to loosen up any dead fur. Then push your fingers in under the fur to work upward from the tail to the head. Use the brush in slow, deep, long strokes from head to tail. If the cat is shedding a lot, use short strokes. With long-haired cats, twist the brush outward, away from the body. Use the same procedure on the tail to fluff it out and to keep the fur from getting tangled. Always follow the line of the fur, and be careful not to brush too hard or vigorously, or you may actually create bald spots in cats that shed continuously. After a thorough brushing along the back and tail, down the legs, and along the sides and belly, stroke the cat with your hands to give the cat's coat a final smoothing polish.

## A Cat's Curse: Coping with Fur Balls

A fur ball in the process of being disgorged is most distressing to cats and to their owners. The whole room shakes with resonant retching as the cat tries to throw up a bolus or two of fur that accumulated in his stomach.

For many cats, this is a weekly ritual. So is cleaning the rug, sofa, or wherever else the slimy fur ball is deposited. It is natural for a cat to regurgitate fur balls, since if fur balls are not evacuated in the feces, they can accumulate in the cat's stomach and the consequences can be serious if they are not eliminated. The cat may starve because the cat cannot put away food in his stomach. A sudden loss of weight coupled with interest in food, but inability to hold food down, are cardinal signs of fur ball accumulation in the cat's stomach. Veterinary treatment is then vital to save the cat's life.

Fur balls sometimes cause intestinal blockage and severe constipation, another complication necessitating immediate veterinary attention.

Not all cats that vomit after eating have a stomach too full for food because of fur balls. They may have internal tumors, or may be allergic to certain food ingredients. Any cat that regularly regurgitates food, with or without fur balls or strands of fur in the vomit, should be taken to the veterinarian for treatment.

This common problem can be prevented in large measure by regularly grooming the cat daily, so that the cat, when grooming herself, swallows less of the fur that adheres to the rasps on her tongue. Your veterinarian can prescribe a mild laxative to help the cat rid herself of ingested accumulations of fur. Also, providing more roughage in the food, such as a high-fiber cat food, or adding some chopped fresh wheat grass, can also help clear the cat's digestive system of swallowed accumulations of fur.

# 22
## | Feline Declawing |

The surgical removal of a cat's claws (often resulting in chronic pain and infection) is often done to prevent scratching of furniture. It cuts out part of what makes a cat a cat. A graphic comparison would be the removal of a person's knuckles.

A declawed cat is less able to protect itself, especially if it is an outdoor cat. It is less dexterous, and far less agile. It is unable to climb trees, leap on owners' shoulders in play, and "go crazy" at night. Some cats also have their hind claws removed, a practice that should be outlawed since this makes the cats even more defenseless (when front claws are also removed) and unable to properly scratch when there is an itch.

Declawed cats often develop chronic lameness due to infection, and, tending to walk more on the back portion of their pads, they develop an abnormal gait that is often compounded by arthritis. Because pawing in the litter box can be painful, especially when infected finger and toe stubs become caked with litter material, some declawed cats develop an aversion to using the litter box. They are then likely to be put up for adoption, especially when the other bad side effect of declawing develops—namely increased irritability and biting. As a source of chronic physical suffering and psychological distress, declawing may bring on stress-related diseases, notably increased susceptibility to infections, allergies, cystitis, and inflammatory bowel syndrome.

The fact that an estimated one quarter of the U.S. cat population—almost fourteen million animals—are routinely declawed is an animal welfare issue that should be addressed by all involved; every measure should be taken to avoid this cruel mutilation. Part of being a cat is to have claws.

With little time and effort on the part of their owners, most cats quickly learn to use a scratching post or board (spiced with catnip), or work out on a cat "condo" with carpeted or rope surfaces.

## More About Declawing Cats

In 2006 the federal government outlawed the declawing of big cats in captivity in the United States, such as lions and tigers, on the grounds of extreme cruelty and suffering. Similar legislation is long overdue for domestic cats who can suffer for their entire lives after such cruel mutilation. My friend Jean Hofve, DVM, has reviewed and written about the harmful consequences of *onychectomy*—the amputation of the digits or first phalanges (not just the claws) of cats.[1] She notes that 33% of cats develop behavior problems after surgery, notably house soiling, biting, and changes in personality; for example, formerly outgoing and friendly cats can become fearful and reclusive.

While laser surgery causes less pain and swelling than other surgical techniques in the first few days after surgery, the long-term consequences of the procedure remain the same. Post-surgical complications include abscess formation, chronic infection (aggravated by cat litter), and chronic or intermittent lameness.

Human amputees often suffer from painful "phantom" limb sensations from the amputated part, a condition that cats may well experience since their nervous systems are virtually identical to ours.

We may not know when some cats are suffering because of their stoic nature. In fact, some cats in great discomfort may actually purr and seem to be half-asleep—such self-comforting, so-called displacement behaviors being indicators of stress. Cats may learn to cope with the chronic pain of onychectomy, but the absence of overt pain does not mean they are pain-free.

The tendons that control the toe joints retract after surgery, these joints essentially becoming "frozen." The toes remain fully contracted for the life of the cat.

I have received a few letters from some cat owners who claim that their cats never developed any problems after being declawed. But I have

---

1. See *Journal of the American Holistic Veterinary Medical Association* (October–December 2006): 34–39.

received many more letters to the contrary, so why run the risk in the first place? There is never any guarantee with this kind of surgery that a cat will not be harmed and endure a lifetime of chronic pain as a permanent cripple. This is why such surgery is illegal or considered extremely inhumane in some twenty-five countries around the world, according to Dr. Hofve, who laments the cavalier attitude of veterinarians performing such surgery in the United States—often without appropriate use of analgesics and advice to cat owners of the possible complications and proper post-surgical care. Surely it is time for the United States to step up to the plate and emulate the European Union's 1987 Convention for the Protection of Pet Animals, which prohibits, for nonmedical purposes, ear cropping, tail docking, declawing, defanging, and devocalizing of companion animals.

# 23
## | Cloning Cats and Dogs |

Goats, sheep, cows, pigs, rabbits, mules, horses, deer, and mice have been cloned for commercial and biomedical purposes. In January 2002, it was announced that the first domestic cat had been cloned. Two of them where exhibited at Madison Square Garden in New York City a couple years later, with an offer by the company to clone people's cats for $50,000 per kitty clone. In August 2005, the first dog was cloned, an Afghan hound, by South Korean researchers at Seoul National University where, earlier, human embryos had been cloned and stem cells extracted. The surrogate mother of this cloned dog was a yellow Labrador retriever. One hundred and twenty-three dogs were used as both egg donors and surrogate mothers; from over a thousand prepared eggs or ova, each containing a skin cell from a dog's ear, three pregnancies resulted—one ending in a miscarriage, one resulting in a pup that died soon after birth from respiratory failure, and the third a viable clone of a male Afghan hound. Some bioethicists fear that the cloning of man's best friend is the final stepping stone to eventual public acceptance of human cloning.

Cloning entails taking a single cell from an animal and placing the cell inside the egg case or ovum taken from another animal of the same species, the egg case having been emptied of its contents. After a procedure that activates the cell to begin to divide, the ovum containing the cloning cell is placed in the uterus of a hormonally receptive surrogate animal. Because of low success rates in getting the cloned cells to implant into the uterine wall, and because the placenta and embryo may not develop normally, several ova containing the clone cells may be put into the surrogate animal's uterus at the same time.

Should the technology be perfected to the point where cloning becomes a business, people will likely be able to take a beloved dog or cat to the veterinarian for a routine health check and have a few cells removed, quickly frozen, and shipped for storage at a pet cloning center. A processing and storage fee will be charged, and when the owners want their companion animals to be cloned, the center will begin the process after a substantial down payment has been made, or full payment has been provided. Before this new biotechnology is perfected and large-scale operations set up with hundreds, possibly thousands, of caged and hormonally manipulated female dogs and cats serving as ova donors, and others being the recipients of ova containing the to-be-cloned pets' cells, the cost will probably be in the six-figure range for some time before mass production follows mass demand. But there are many concerns other than financial.

The cloned dogs and cats will not be exact replicas of people's beloved animal companions, and many clones will probably be spontaneously aborted or destroyed because of various birth defects. Abnormalities may also develop later in life. Clones of other species often have abnormal internal organs, and neurological and immunological problems. They may also be abnormally large at birth due to a defective growth-regulating gene function. What about the origins, quality of life, and future of the thousands of caged female dogs and cats who will be exploited by the pet cloning industry; what about the procedural risks to their health and overall welfare? Do the ends justify the means?

There is no evident benefit to the animals themselves. Why not adopt a dog or cat who needs a good home from an animal shelter? Or donate money, equivalent to what it would cost to produce one clone, toward improving the welfare of hundreds—even thousands—of dogs, cats, and other animals in communities around the world?

What are these ends anyway? Certainly there is a commercial end that is potentially lucrative, given the right market promotion and endorsements by professionals and celebrities. But is there real human benefit in making a clone of one's beloved animal companion? Or is it mere pandering to a misguided sentimentalism? Because of the close emotional bond between humans and their animal companions, the pet cloning business can be seen as an *unethical exploitation of the bond* for pecuniary ends. Exact replicas of people's dogs and cats will not be created because an identical environment during embryonic and postnatal development cannot be achieved. All clones may, at the time of birth, be of the same chronological age as the age of the cells taken from the to-be-cloned animals. So if a cell is taken from a six-year-old dog, because of the aging "clock," the clone may already be aged by six years at the time it is born.

From various religious and spiritual perspectives and beliefs, cloning violates the sanctity of life and the integrity of divine or natural creative processes. It is problematic from the point of view of reincarnation, or transmigration of the soul. From a Buddhist perspective, the consciousness incarnate in the body of the clone, or the consciousnesses in the bodies of many clones from the same original animal, are all going to be different from the original donor.

It is not inconceivable that dog and cat clones might also be created initially on an experimental basis and used to provide spare parts such as kidneys, hearts, hips, and knees for ailing dogs and cats. Research laboratories may also use cloning to quickly develop identical sets of dogs, cats, and other animals for biomedical research. Some sets and lines of clones having the same genetically engineered anomalies to serve as high-fidelity models of various human diseases may be created and marketed to develop new and profitable drugs to treat these conditions in humans and other animals.

The bioethics and medical validity of these developments need to be examined. And pet owners who put out the money to have their animal companions cloned may want to think twice, since they may well be giving this new cloning business not only a financial jump start, but also the socio-political credibility that it needs in order to gain widespread public acceptance and a market for human cloning and other biologically anomalous and ethically dubious products and processes.

The fact that a venture capitalist made a grant of $2.3 million and hired an agent to find a university biotech laboratory already in the cloning business to clone his dog Missy (visit www.missyplicity.com), and the subsequent public relations and media promotion of this project, points to another agenda: the cloning of pets may be a ploy to promote human cloning. If the cloning of pets becomes a reality, the public will become desensitized to the issue of cloning and more likely to eventually accept a highly lucrative biotechnology for childless couples and rich and selfish singles with a desire to clone complete human beings or partial human beings (such as anencephalics or headless clones) as a source of replacement tissues and organ parts.

The Philosophy Department at Texas A&M University—where the Missyplicity Project was started in another department before being spun off into a private company called Genetic Savings & Clone—developed a set of "bioethical guidelines" based on the ethical principle of what they call *axiomatic anthropocentrism*. This strategy was clearly designed to deflect public criticism and concern over the morality and animal welfare aspects of the project.

Axiomatic anthropocentrism essentially means that whatever is good for people is ethically acceptable. Anthropocentrism—human centered-ness—is an outmoded worldview or paradigm that many advocates of animal rights and environmental protection see as the root cause of untold animal suffering and ecological devastation over the millennia.

Several female dogs were up for adoption on the Web site, one of the company's "bioethical principles" being regardless of the source through which dogs are obtained for use as egg donors or surrogate mothers (from animal shelters, breeding farms, etc.), at the completion of their role in the Missyplicity Project, all dogs shall be placed in loving homes. No funds shall be expended for dogs raised under inhumane conditions, such as puppy mills. The Missyplicity Project included several goals in addition to the cloning of Missy that have been published on the Web site. These included dozens, perhaps hundreds, of scientific papers on canine reproductive physiology; enhanced reproduction and repopulation of endangered wild canids; plans to develop improved canine contraceptive and sterilization methods as a way of preventing the millions of unwanted dogs that are euthanized in America every year; plans to clone exceptional dogs of high societal value, especially search-and-rescue dogs; and to develop low-cost commercial dog-cloning services for the general public.

These goals gave the project the kind of credibility that a gullible public, as well as organizations and professionals with a limited grasp on the inherent limitations and harmful consequences of cloning, would readily accept. Ethical concerns and the questions concerning the validity and relevance of applying cloning biotechnology to wildlife conservation, to dog overpopulation, and to the propagation of high-performance dogs were cleverly deflected by these promissory goals.

The veterinary profession has been relatively silent on this issue of the risks and ethics of companion animal cloning. I trust that my respected colleagues will not remain silent on this issue, as they did thirty years ago with the advent of factory farming that has resulted in great animal suffering, environmental harms, and increased public health risks.

# 24
# | The Euthanasia Question |

When a fellow being is suffering without hope of relief and recovery, those who care will opt for mercy killing (euthanasia). Euthanasia is unacceptable in some cultures and religious traditions because to kill another being deliberately, such as a holy cow in India, is taboo. This has less to do with the life and plight of the animal than with the shame of making oneself "impure" by killing another. Self-interest, in this instance, takes precedence over compassionate action. Such inaction (letting the animal continue to suffer) is more than cowardice. It is the essence of hypocrisy when euthanasia is equated not with compassion but with violence, disobedience to some religious doctrine (like ahimsa, nonharming), and personal defilement.

The attitudes in the West toward euthanasia are no less divided and divisive. While the euthanasia of animals is culturally accepted, there is no consensus over the euthanasia of people, even of the terminally ill who are able to give informed consent to caring doctors, and who have supportive relatives.

Any cultural or religious moral principle such as ahimsa, or respect for life, cannot be absolute since there are extenuating circumstances (i.e., situational ethics). Killing an animal out of compassion can be a purely selfish choice. I have been in situations where my decision to euthanize an animal was difficult to separate from my own empathic suffering. Putting the animal out of her misery would put an end to my own burden of suffering for the animal. It is difficult in some situations to be compassionate and at the same time sufficiently detached to be able to make the right decision, free of self-interest, especially when facing the decision to have a beloved companion animal euthanized and one is suffering with it. Empathy, and sympathetic identification, must be balanced by clinical objectivity and assessment of the situation in terms of the animal's degree of suffering and chances of recovery, and options for immediate, short-term relief, as with analgesics and tranquillizers.

I recall one instance in India, treating a Pomeranian dog named Snowflake who had lost most of the skin on her back (about one-third of her surface area) from an accidental scalding by her owner. My first reaction, while cleaning the massive wound, was to consider euthanizing her. But strangely, she seemed to be in less pain than I had anticipated, and with her indefatigable spirit and will to live, combined with intensive wound care and love, she healed completely in three months.

So it is important when considering if it is time to have one's companion animal euthanized to have a second opinion from a veterinarian, and ideally an impartial third-party opinion, as from a close friend who understands your connection with your animal companion but can be more impartial than you. Some people will hold on for emotional reasons, or out of some irrational hope that the animal may miraculously recover while in fact the animal's quality of life calls for letting go. Terminally ill animals in the wild will often seek solitude and a safe place to die, just as companion animals may become more withdrawn and inward, even going off if they get outside to hide before they die.

An animal who shows less and less interest in life, in food, water, daily activities, especially the routine walk or playtime and petting time, and is either more withdrawn or restless, even agitated and seeking extra attention—all signs of pain, fear, and what I call *senile dysphoria*, the antithesis of euphoria—should be seen to immediately by a veterinarian. Ideally a house-call by the veterinarian may be preferable to taking the animal to the hospital or clinic—unless the animal is very easygoing, was never unduly upset going to see the veterinarian on prior occasions, or is so out of it that a short car ride cradled in your lap would not cause the animal undue distress. Some animals do seem to know when it is "their time," and animals sharing the same home will often react by being more attentive and solicitous, or fearful and avoid contact, much like people do toward their own kind.

For many animals, in-home euthanasia is the most humane. It may be best for some family members, including animals not to be present during the procedure. But viewing the body after euthanasia may be an important closure for all concerned. Likewise at the veterinary hospital, some veterinarians do not want the animal's owner involved, but in other instances one or more human family members may be allowed into the room and help hold the animal while the procedure is undertaken.

One of the more widely adopted procedures is to first inject the animal in the thigh or other deep muscle area with a tranquillizer prior to giving an intravenous injection containing a barbiturate drug that causes rapid unconsciousness, like a general anesthetic, but from which the animal never recovers because an overdose is given, sometimes in combination with other drugs to stop the heart. As the animal passes rapidly to unconsciousness, body movements, muscle spasms, changes in breathing, including gasping, sometimes with a vocal sigh or moan, may occur. These can be extremely upsetting to the uninformed but are body reactions associated with the brain and circulation shutting down after the animal has lost consciousness and is not, therefore, suffering. Some animals simply collapse with a short sigh, as though in relief moments after the intravenous injection is administered. But because of the reactions, which can never be predicted, many veterinarians prefer not to have many of their clients witness the procedure.

Death with dignity and freedom from fear are the goals of euthanasia. The grief, guilt, and tendency to blame oneself or others for the loss of an animal companion are all part and parcel of the mourning process. Seeking the support of other family members and friends, attending support groups and grief counseling sessions, often available through veterinary referral or at your local humane society, can be well advised. The grieving over the loss of an animal companion can be unexpectedly painful and enduring, complicated by depression and despair that could even be life-threatening. Rather than focusing on the loss, on blame and shame, one needs instead to remember the good times—the unconditional love and joy the animal brought into one's life, and the final gift to the beloved animal of providing a humane death that liberated the animal's spirit from a worn-out body, preventing further pain and fear. This is all part of the recovery phase of letting go.

In the final analysis we should adopt neither a sanguine nor an abolitionist attitude toward euthanasia of animals—or of our own kind for that matter. The compassionate middle-ground between these extremes can be difficult to establish, as for instance in some non-Western countries where appeals to reason and compassion with regard to animal euthanasia can evoke violent opposition. It is indeed tragic when abandoned cows starve to death and when homeless dogs are neutered and released, only to suffer a hopeless existence on busy city streets before they are killed by traffic or are rounded up by municipal dog catchers to be killed by electrocution or with injections of strychnine and cyanide. The Western humanitarian policy of humanely euthanizing unadopted homeless dogs is anathema to many of their counterparts in the East. Surrounded by suffering, people can become desensitized to it. The attitude of live and let live can then have cruel consequences when responsible euthanasia is taboo—just as when, for purely selfish reasons, the inability to let go and let die leads to taking extreme measures to keep a loved one alive, regardless of the individual's quality of life and suffering.

# 25
## | What Makes Animals Happy? |

What makes some people happy can make others sad and angry. Take, for instance, a Minnesota hunter who filed a hunter harassment suit on grounds that the defendant spoiled his pleasure in hunting and shooting a bear. The defendant, a renowned bear field biologist, had verbally abused him when the hunter, still in his tree hide, started to ridicule the bear biologist's graduate student as she wept over the dead body of the black bear she had studied for over a year—an animal she had come to know and love.

What makes me happy is assuring that the happiness of others, both human and nonhuman, are not mutually exclusive but mutually inclusive and enhancing. The happiness of one should not impinge upon the happiness of another. That, surely, is the essence of a truly democratic society based on the bioethics of compassion and equalitarianism, rather than on a utilitarian one that would sacrifice the happiness of a few for the happiness of the many. A happy companion animal makes for a fulfilling relationship with the human care-giver and family, just as happy farm animals mean healthier and more productive animals and more healthful produce.

For the sake of argument one might say, OK, then for the happiness of deer we should prohibit all hunting, even of deer by wolves and other wild predators. For the happiness of house cats we should buy mice for them to enjoy killing. But not buying mice for kitty to kill will not detract from a well-cared-for cat's happiness and overall well-being. The same can be said for a man who feels that his right to kill bears for pleasure takes precedence over the happiness of others—including those who would like to see an end to all nonsubsistence "sport" or recreational hunting and killing, and those who love bears for bears' sake and not as hunting trophies.

As for the relationship between deer and wolves, theirs is a mutually enhancing symbiosis because the deer herds are kept healthy and in balance with the ecology, thanks in part to wolf "management." Also, the deer do not go around in terror expecting to be killed, but enjoy a quality of life, and happiness indeed, in the wild where they have close social bonds, play together, and care for their offspring with devotion equal to any human.

Significantly, the British government has formulated new animal welfare regulations that stipulate a "duty of care" for all people who keep animals as companions/pets. The well-being of animals is equated with the provision of basic freedoms—a point underscored by the British Veterinary Association's Animal Welfare Foundation's www.bva-awf.org.uk educational brochure entitled "What Makes My Pet Happy?" introduced at the beginning of this book (see chapter 1). Again, the basic Five Freedoms that provide for animals' physical and mental health are: freedom from hunger and thirst; freedom from pain, injury, and disease; freedom from discomfort (e.g., temperature, floor surface); freedom to express normal behavior; and freedom from fear and distress. In the association's brochure, after asking if a dog left alone all day or a budgie sitting alone in a cage are actually happy, the text states: "Happiness, welfare and quality of life are all talking about how animals feel."

Happiness can be defined as a subjective state of well-being that includes emotional security and physical contentment/satisfaction. It is clearly evident in animals' playfulness, friendliness, and displays of affection and trust toward their care-givers.

But no matter how good the quality of care may be, an animal's happiness and well-being can be undermined by a lack of adequate socialization or impaired bonding with their own kind and/or with humans during their formative early weeks of life. Genetic factors can also play a beneficial or detrimental role; hereditary factors can be associated with fear of strangers (especially in captive wild animals and domestic animals who are innately more fearful than normal) or with an outgoing, curious, and stable temperament—meaning that some animals live in fear, while others enjoy a happier existence.

With both fearful domesticated animals and captive wildlife, much can be done to improve their quality of life through sensitive handling and provision of a secure and comfortable environment—ideally with the supportive companionship of other animals that are easygoing and emotionally stable.

But as prevention is the first medicine, chronic unhappiness in fearful animals, both wild and domesticated, can be respectively ameliorated by establishing a close bond with the caretaker/keeper (which is not without some human risk when it comes to species like elephants and tigers) and by environmental enrichment, which means providing as much natural habitat conditions as possible. More rigorous selective breeding of domesticated animals for adaptability and stability of temperament—by not breeding excessively shy, timid, and fearful or aggressive individuals—can also do much to improve their overall well-being.

The fourth of the Five Freedoms—freedom to express normal behavior—is an integral aspect of an animal's living conditions. It continues to be denied, primarily for reasons of cost, to animals raised for human consumption in factory-farm cages and pens and feedlot enclosures, including those animals raised for their fur pelts. It is also denied to the following animals: those animals—especially lions, tigers, elephants, and bears—who are exploited by the circus industry; wildlife held captive in roadside menageries, marine-life aquaria, and Third World zoos; and most species kept in small laboratory cages for experimentation and product tests. The solitary, impoverished existence endured by millions of small "pets"—especially highly social rabbits, mice, and other rodents, and various other species including "exotic" mammals, birds, reptiles, and amphibians—needs to be addressed from the standpoint of animals being denied the freedom to engage in normal social behaviors with their own kind, and with a duty of care that provides an environment and quality of life conducive to the animals' happiness.

Quality of life matters to humans just as it does to all other animals. Professor K. Hartmann of the Ludwig-Maximillians University, Animal Medical Clinic (Munich, Germany) sent me an effective clinical adaptation of the Karnofsky Quality of Life scoring protocol, used for determining quality of life for human lung cancer victims, adapted for cats. He and his colleague, Professor M. Kuffer, treat sick cats with feline AIDS (feline immunodeficiency virus) on a regular basis.

He developed a detailed quality-of-life questionnaire for cat owners to answer—including such questions as to their cats' playfulness, sleeping and eating patterns, toilet and grooming behavior, and overall interest in life. Establishing a baseline norm, he was able to better determine the effectiveness of two antiviral compounds (PMEA and FPMA) in making AIDS-infected cats feel better.

What Prof. Hartmann and his colleagues have demonstrated is that quality of life can be quantitatively and objectively determined in animals with a little effort. This takes the sound science of applied animal ethology and clinical behavioral assessment—and not a stretch of the imagination, as some might contend, believing that quality of an animal's life (and therefore happiness), unlike that of a human, cannot be scientifically verified. Proof of clinical improvement, as observed by Profs. Hartmann and Kuffer, can be objectively determined by close observation of cats' behavior, which in turn reflects quality of life, health, and well-being.

Many of my veterinary and animal behavior research friends and colleagues, some of whose fine work I have cited and referenced in this book, have done a great service to advancing our understanding and appreciation of animals by documenting countless examples of how and why animals seek pleasure, show happiness, and express joy—and how they suffer, especially from our inhumanity. The more we all ponder the question of what makes animals happy, and discover how we can make them happy—in both body and mind—the better we will meet our responsibilities to care for all creatures under our dominion. Their well-being is a duty that every society that is civilized embraces to the full without distinction or exception. The measure of civilization's progress, and of our humanity, is how well other animals are, be they companion cats, wild cats great and small, or other creatures whose fate is tied to our own destiny on this fragile and beautiful planet that we share with all.

I cannot close without a personal word of deep appreciation for all the animals who have enriched my life in countless ways, like Igor, my first cat. They have enabled me to share in this book something of their wisdom and living presence that not only reflects the beauty and mystery of life, but also deeper realities like the empathosphere, of which we are all a part, in body, mind, and spirit.

# Postscript

The company Genetic Savings & Clone, a commercial spin-off from the Missyplicity Project at Texas A&M University, launched Operation Copy-Cat in 2000. The company estimated that the price for cloning a cat or dog will drop to $25,000 within three years.[1] But because of their failure to clone a dog, and lack of interest in people having their cats cloned, they closed shop, sending stored dog and cat tissues to another holding company to be kept in cold storage in order to appease their customers who still hope that their pets may someday be cloned.

---

1. For additional information about genetic engineering, cloning, and the creation of transgenic animals see: Michael W. Fox, *Killer Foods: When Scientists Manipulate Genes, Better Is Not Always Best* (Guilford, Conn.: Lyons Press, 2004); Michael W. Fox, *Bringing Life to Ethics: Global Bioethics for a Humane Society* (Albany, N.Y.: State University of New York Press, 2001).

# | Index |

Italicized page references indicate photographs. Footnotes are indicated with "n" following the page number.